Lecture Notes in Mobility

Series editor

Gereon Meyer, Berlin, Germany

More information about this series at http://www.springer.com/series/11573

Ingemar Denbratt · Aleksandar Subic
Jörg Wellnitz
Editors

Sustainable Automotive Technologies 2014

Proceedings of the 6th ICSAT

Springer

Editors
Ingemar Denbratt
Applied Mechanics
Chalmers University of Technology
Goteborg
Sweden

Jörg Wellnitz
Faculty of Mechanical Engineering
Technische Hochschule Ingolstadt
Ingolstadt
Germany

Aleksandar Subic
School of Aerospace, Mechanical and Manufacturing Engineering
RMIT University
Bundoora
Australia

ISSN 2196-5544 ISSN 2196-5552 (electronic)
Lecture Notes in Mobility
ISBN 978-3-319-17998-8 ISBN 978-3-319-17999-5 (eBook)
DOI 10.1007/978-3-319-17999-5

Library of Congress Control Number: 2015939676

Springer Cham Heidelberg New York Dordrecht London

Printed on acid-free paper

Springer International Publishing AG Switzerland is part of Springer Science+Business Media (www.springer.com)

Preface

The world's population is growing rapidly; it is expected to increase from 7 billion people today to between 8 and 9 billion in 2030. In the same time frame, the global road fleet will grow from 1 billion vehicles today to an expected 1.7 billion. This, together with rising gross national products, will push energy consumption to ever higher levels, causing further increases in atmospheric CO_2 levels.

Moreover, air pollution is becoming a serious issue in megacities. There is a broad consensus that the burning of fossil fuels must be reduced in the future, mainly by increasing energy efficiency but also by gradually increasing the contribution of renewable fuels to the global energy mix.

There are many factors that will affect future developments in transport and energy use, including the availability and cost of fuel; regulatory standards governing fuel economy, exhaust emissions, and the use of renewable fuels; technological advances; globalization and urbanization (especially changes in the way vehicles are used and the organization of transportation hubs); infrastructure; and economic growth, especially in emerging and developing markets.

The automotive industry is undoubtedly a major driving force for much of the technological development that occurs in modern society, and this is likely to remain the case for the foreseeable future.

In the medium term, research and development activities relating to cars, trucks, and buses will be primarily focused on improving engine and vehicle efficiency, while maintaining compliance with increasingly stringent exhaust emissions standards.

Key areas of vehicle technology that will be targeted to improve fuel efficiency between 2015 and 2025 include propulsion systems, overall vehicle design and aerodynamics, materials, electrification, parasitic losses, heat recovery, and integration.

While many of the new technologies that will be needed have already been tested and demonstrated in principle, their details remain to be worked out and they will have to be integrated in a holistic way. This will necessitate the development of new control systems and methods for optimizing overall vehicle efficiency.

This book on Sustainable "Automotive Technologies" describes a cross-section of current international research activities aiming to address the challenges that will be presented by both the market and regulatory authorities in the near future.

The book comprises six main chapters; (i) Electromobility, (ii) Systems solutions for sustainable mobility, (iii) Combustion engines, (iv) Materials and Light Weight Design, (v) Alternative and renewable fuels, and (vi) Sustainability. Papers selected for publication in this book have been presented at the 6th International Conference on Sustainable Automotive Technologies, ICSAT2014, held in Göteborg, Sweden in September 2014.

All of the contributions included in this book have been reviewed independently by international experts with experience in relevant fields and have been edited accordingly prior to publication.

We wish to take this opportunity to thank all of the participating researchers and reviewers for their contributions, without which it would have been impossible to produce this book. We also gratefully acknowledge Springer's continuing support of this research field, which has enabled us to establish and publish an international book series on Sustainable Automotive Technologies that has attracted worldwide interest.

Finally, special thanks are due to Elenor Norberg for her brilliant work in organizing ICSAT 2014.

Ingemar Denbratt
Aleksandar Subic
Jörg Wellnitz

Contents

Part I
Electromobility

Range Extender Engines for Vehicles with Electrical Propulsion

F. Ekström and J. Engman

Abstract In recent years, electrified vehicles have gained renewed interest. However, one of the large hurdles for vehicles with electrical propulsion is the battery capacity and thus the limited driving range. Thus, some kind of range extender has been proposed, i.e. a unit transforming liquid fuel into electricity and/or mechanical energy for extended driving range. Here we present an approach to determine the size and power of a range extender, as well as the methodology of multidisciplinary optimization to optimize the vehicle system attributes, leading to three different hybrid range extender vehicles, which have been built and tested. We conclude that the most beneficial system from a cost/performance point of view as well as the lowest fuel consumption can be found using a mechanical link between the internal combustion engine and the traction wheels.

1 Introduction

The first hybrid vehicle was designed more than a century ago by Ferdinand Porsche, the Porsche-Lohner Mixte, also known as the first all-wheel drive vehicle, featuring wheel hub motors and a series hybrid powertrain [1]. However, the concept was soon abandoned due to its high fuel consumption, while the inventor explored new tracks for mobility with well-known results.

Volvo Car Corporation has a long history of electrification projects, also shown externally, such as EEC (1992, Series hybrid), HEV98 (1998, Series hybrid), Desiree (1999, Power Split hybrid) [2]. Lately, this work has gained renewed interest due to upcoming CO_2-emission legislation, heading for 95 g CO_2/km fleet average in EU from 2021. Electrified vehicles may become a part of the solution to reach these goals. However, since pure battery electric vehicles require heavy, bulky and expensive batteries to achieve an acceptable driving range, some kind of

F. Ekström (✉) · J. Engman
Volvo Car Corporation, Gothenburg, Sweden
e-mail: fredrik.ekstrom@volvocars.com

I. Denbratt et al. (eds.), *Sustainable Automotive Technologies 2014*,
Lecture Notes in Mobility, DOI 10.1007/978-3-319-17999-5_1

range extender unit might be necessary. In recent years, a number of different range extender concepts have been presented [3–8]. Although a range extender can be a thermodynamic machine of many different kinds, such as 2-stroke engines, Wankel engines, Stirling engines, steam engines, gas turbines, free piston engines or even fuel cells, the requirements for usage in an automotive application limits the number of options substantially. These requirements are in particular packaging space, weight, NVH, emissions, production cost, vehicle integration cost, development cost and risk as well as good startability. Due to these aspects, the 4-stroke Otto-engine stand out as a proven technology, passing all requirements at a reasonable cost.

2 Hybrid Topology

Our definition of a range extender is a fuel converter unit in a powertrain where the main vehicle traction force comes from an electric machine. A simple overview of two different range extender solutions can be found in Fig. 1, showing examples of a series hybrid installation and a parallel hybrid installation.

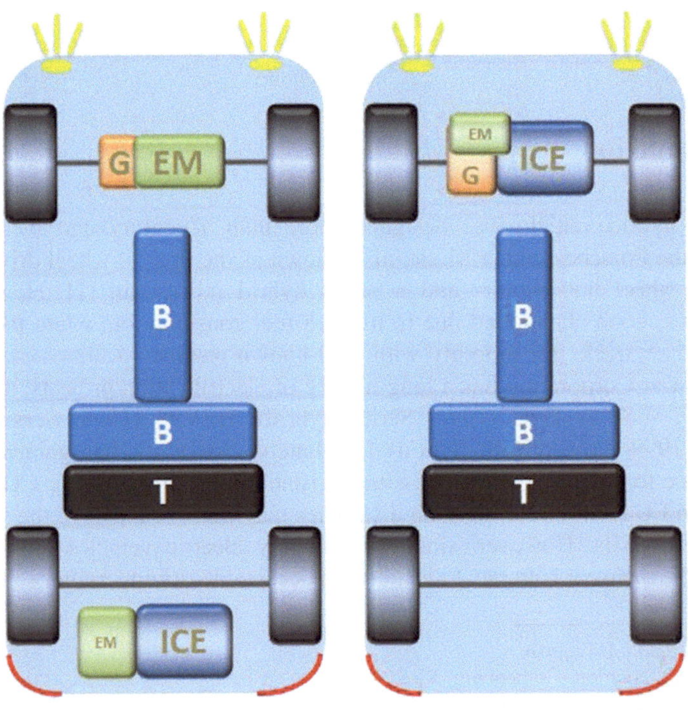

Fig. 1 Range extender for series hybrid solution (*left*) and parallel hybrid solution (*right*)

Table 1 Typical efficiencies for the electrical system in a series hybrid powertrain

Efficiency	Generator	PCU	Battery	Motor
Value	0.8–0.96	0.95	0.9–0.95	0.8–0.96

In a series hybrid (SH) installation, the traction force to the wheels comes from the electric machine only, via a transmission with one or several gears. Thus, the vehicle is always propelled by electricity, and the only function of the range extender is to create electrical energy for the battery and the electric machine via a generator, i.e. the series hybrid solution requires two electric machines.

In a parallel hybrid installation (PH), both the electric machine and the range extender engine can propel the wheels, i.e. there is a mechanical linkage between engine and wheels, and the power of the electric machine and engine can be combined for propulsion of the vehicle. Also, the generator is omitted.

Comparing the two range extender system solutions at constant vehicle speed, we obtain the following system efficiencies:

$$\text{System efficiency}_{SH} = \eta_{ICE} \cdot \eta_{Generator} \cdot \eta_{PCU} \cdot \eta_{Battery,charge} \cdot \eta_{Battery,discharge} \cdot \eta_{PCU} \cdot \eta_{Electric\ Motor} \cdot \eta_{Transmission} \tag{1}$$

$$\text{System efficiency}_{PH} = \eta_{ICE} \cdot \eta_{Transmission} \tag{2}$$

Thus, the efficiency of the electrical system is a key to low fuel consumption of the series hybrid concept. Typical values of these efficiencies, which vary with operating point, can be found in Table 1.

3 Analysis of the Range Extender Power Requirement

The main difference between an internal combustion engine of a conventional drivetrain and a range extender engine in an electrified powertrain is that a conventional engine must deliver a certain torque to provide the vehicle with traction force, i.e. acceleration, while the range extender engine only delivers the average power need of the vehicle, since the traction force is primarily accomplished by the electric motor. For a conventional powertrain traction force is created by the available engine torque and gear ratios between engine and wheels, but it is inevitable that the engine by this method will be mostly working in the partial load range, see Fig. 2, and thus always have a torque reserve for acceleration available.

However, for a reasonable downsized engine and a fuel economy oriented gearing, the operating area of the highest gear is normally within the area of the lowest specific fuel consumption of the engine. The vehicle speed where this occurs is usually above 100 km/h, i.e. the highest efficiency of the powertrain is paired with the highest power demand of the vehicle.

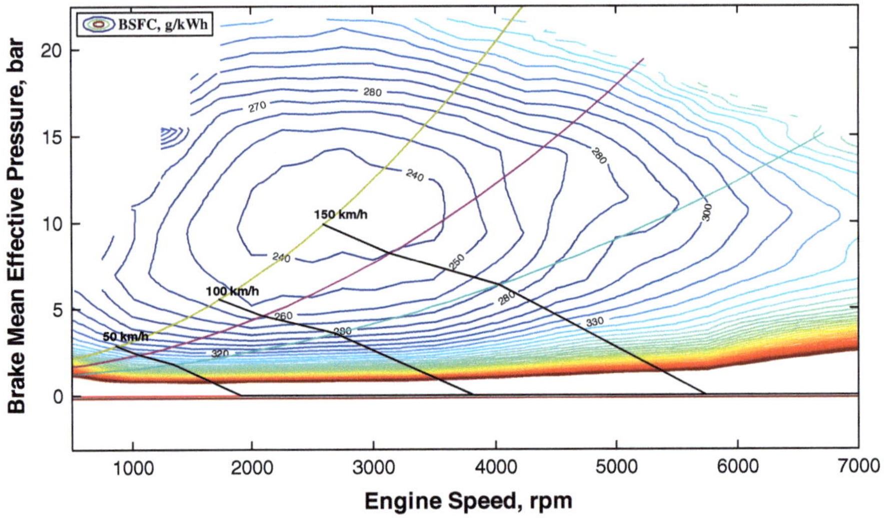

Fig. 2 Specific fuel consumption for a 2.0 L turbocharged Otto-engine, and the associated operating points for a D-segment vehicle at different gears

To clarify the power needs of the vehicle on an average, and thereby the size and maximum power of the range extender engine, it is advantageous to study how vehicles have been driven in various traffic situations [9]. A common distinction is to separate these traffic situations in city driving, rural and motorway driving, see Table 1, where also the power requirements of the vehicle at possible continuous top speeds have been included.

Additional criteria must be added to the values of Table 2 on vehicle level, i.e. auxiliary power needs such as air conditioning, power steering and onboard power consumption, as well as external factors such as hilly terrain, road curvature, high altitude and trailer load. However, most important in the sizing of the range extender engine is the inclusion of the inefficiencies according to Eqs. (1) and (2). This was performed in a vehicle simulation program for the transient driving cycles of Table 2, and the corresponding average range extender engine power needs versus average vehicle speed are shown in Fig. 3.

From the power requirements in Fig. 3, there are at least two ways to control the combustion engine in a series hybrid installation:

1. Always operate the engine at the lowest Brake Specific Fuel Consumption
2. Control the engine power output according to the current vehicle power requirement, i.e. a so called load following strategy

By selecting control method 1, the engine size is chosen in order to achieve the best specific fuel consumption in an operating area frequently used by the vehicle, while at lower power requirements, the engine is switched off and on. At higher power requirements, the engine speed is increased. Using this control method, the

Table 2 Power and energy requirement at the wheels for a vehicle in the D-segment for various driving cycles

Cycle	Average speed (km/h)	Propulsive energy (Wh/km)	Regenerable energy (Wh/km)	Net energy (Wh/km)	Average power need (kW)
Artemis urban	17.6	189	−123.1	66.2	2.25
Artemis rural	61.4	153	−37.5	115.9	8.3
Artemis motorway	120.3	220	−14.4	205.5	25.6
Top speed, 130 km/h	130	221	0	221	28.7
Top speed, 160 km/h	160	304	0	304	48.7

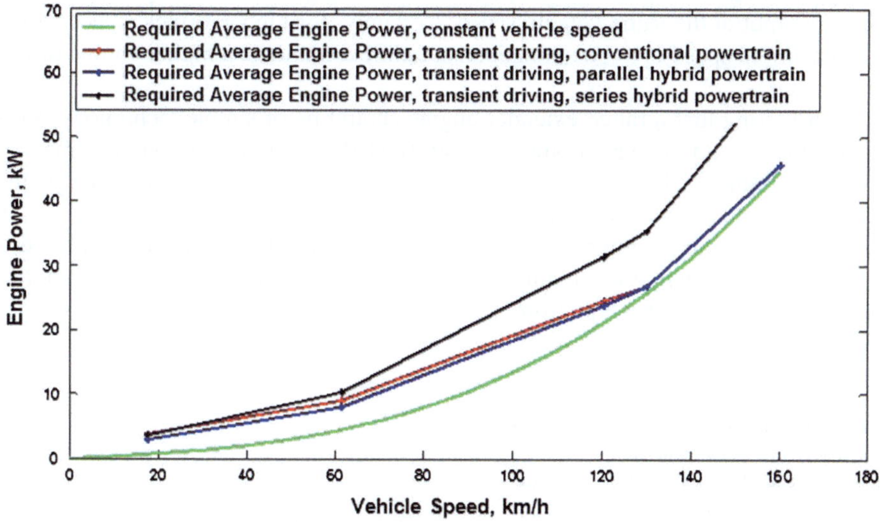

Fig. 3 Power requirement for the engine at constant vehicle speed and average engine power requirement for transient driving cycles versus average vehicle speed

BSFC of the engine will be low, but since the power is not used directly, a large amount of energy will be stored in the battery at lower power outputs, followed by large losses, see Table 1. Also to be considered is the sound quality inside the vehicle, with more or less one constant engine speed, independent of vehicle speed [10–12]. Another aspect to be considered is the cooling need of the engine. At high engine power output and low vehicle speeds, forced cooling by fans is required, which adds to the auxiliary power needs and thereby associated parasitic losses.

With control method 2, the range extender engine speed is primarily used for power control, i.e. a virtual gear is created, where the engine speed follows the vehicle speed. As can be seen in Fig. 3, the required range extender power on average shows a fairly linear behavior vs vehicle speed up to around 130 km/h, i.e. the engine can be run on high loads with primarily engine speed as power control. In this way, the virtual gear creates a natural connection between engine speed and vehicle speed, too high charging losses through the battery is avoided and the cooling of the combustion engine can be handled through headwind, i.e. without cooling fans and the thereby associated losses.

The upper power limit is determined by the maximum speed acceptable for interior noise levels. At highway speed, we believe this engine speed is around 4000 rpm, which gives an acceptable fuel penalty due to higher friction losses at higher engine speeds.

Thus, from Fig. 3 the necessary power of a range extender engine can be deducted as a function of maximum vehicle speed. For this example, a vehicle in the D-segment, around 36 kW continuous engine power output is necessary for a series hybrid at 130 km/h, while a parallel hybrid range extender only needs 27 kW power output at the same vehicle speed. Certainly, for higher vehicle speeds, or if the maximum speed is to be maintained at uphill conditions, a higher engine power is necessary.

Considering that a range extender engine should be of simple technology, such an engine should deliver around 11 bar BMEP of torque without external air charging. Figure 4 shows the corresponding engine power for this torque level, and if we choose to run the engine at 4000 rpm at 130 km/h, it can be concluded that an engine size of around 1 liter displacement is suitable. For this engine displacement a 2- or 3-cylinder engine is suitable, where a 2-cylinder should be slightly better in fuel consumption, but worse from a noise, vibration and comfort perspective.

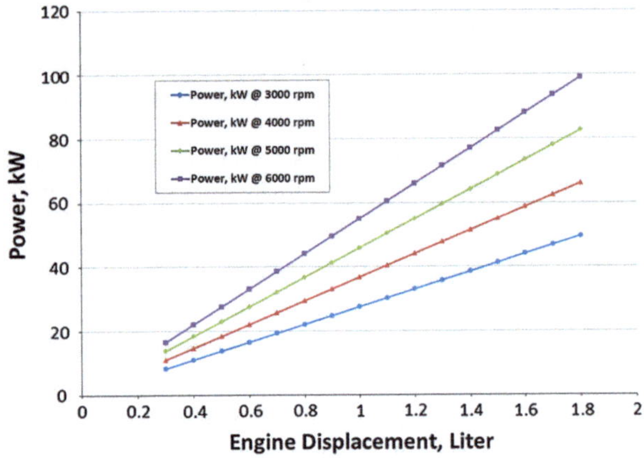

Fig. 4 Engine power as a function of displacement and engine speed at 11 bar Brake Mean Effective Pressure (BMEP)

4 Complete Powertrain Analysis and Choice of Concept

In order to make concept choices for a complete powertrain and electrified vehicle, a multidisciplinary optimization (MDO) [13] was performed (Fig. 5). For this purpose, simulation models were created for the individual components of the powertrain and the variation within each component, i.e. combustion engine (displacement, number of cylinders), electric machine (size and power), battery (capacity), transmission (type and gear ratios). In this way, thousands of powertrain concepts could be simulated.

The simulation results, used as objective functions for the optimization, were mainly fuel consumption, performance, electric vehicle performance and product cost. These attributes were weighted according to customer acceptance criteria and finally multiplied to create a total attribute for this product type as the main objective function, which thus serves as a measure how well the product would be accepted by customers.

The objective function, i.e. the Total Attribute Function (Fig. 5), plotted against total powertrain cost, thus serves as the measure for best cost benefit ratio.

Following the MDO analysis, three different concepts were chosen, which have been built and tested [14, 15] (Figs. 6, 7 and 8):

Concept I. Range-Extender in a series hybrid installation. The vehicle is a C30 Electric, with the propulsion electric motor in the front, and the range extender and the generator located in the rear under the cargo floor.

Concept II. Range-Extender in a parallel installation. This concept is the same vehicle as in concept I, but the generator of the range extender is replaced by a transmission and a second electric traction motor. In this way, all wheel drive can

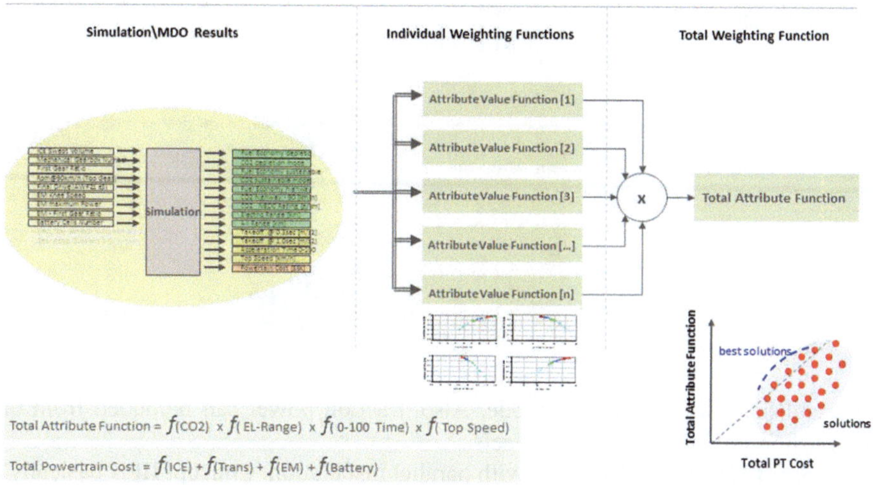

Fig. 5 Multidisciplinary optimization as a method for concept choice

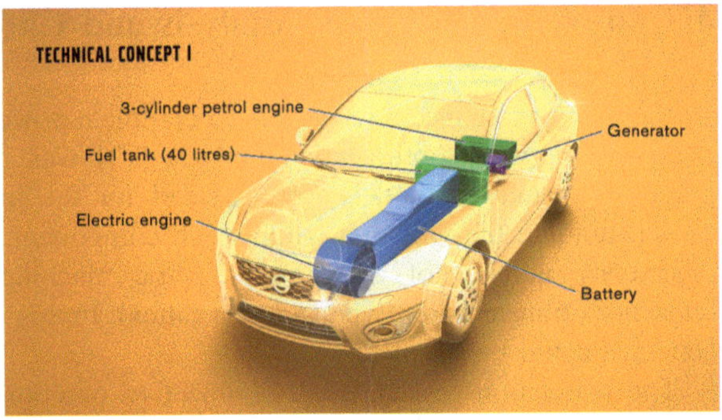

Fig. 6 Concept I

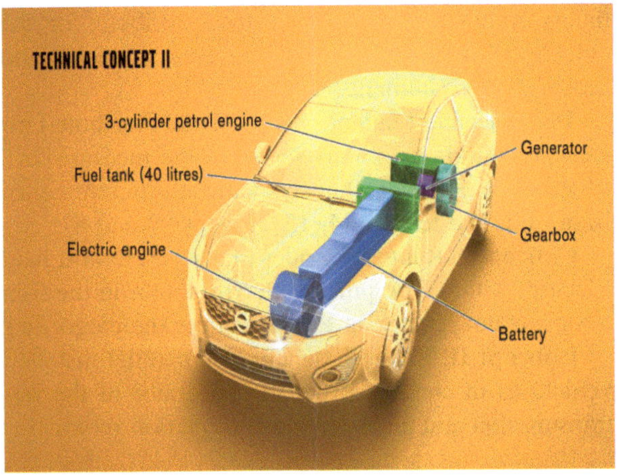

Fig. 7 Concept II

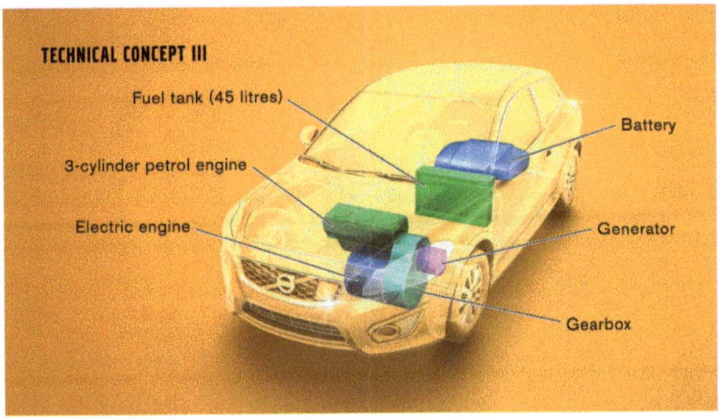

Fig. 8 Concept III

be established in pure electric mode. Also, traction power can be added from the combustion engine to the rear wheels.

Concept III. Range-Extender with parallel installation. Concept III is basically a V60 plug-in hybrid vehicle (PHEV), where the 5-cylinder diesel engine is replaced by a 3-cylinder Otto-engine. Additionally, the original rear wheel electric motor has been replaced by a more powerful electric motor fully integrated in the front transmission, in order to enhance the perception of an electric vehicle.

5 Conclusions

Electric vehicles can be equipped with range extender for prolonged operating range, and the currently most optimal device for this is a standard automotive style 4-stroke Otto-engine. The size and power of such an engine can be determined through analysis of driving cycles of different pattern. The usage of a series hybrid powertrain increases the necessary power output by up to 30 % at highway speed, which is difficult to compensate for by increased efficiency of such an engine. Furthermore, complete powertrain optimization is necessary for electrified vehicles. Multidisciplinary optimization has been effective in defining the complete power-train layout. Following the analysis, a range extender electric vehicle has the highest fuel efficiency and the best cost benefit ratio using a mechanical link between combustion engine and the traction wheels.

References

1. Wakefield E (1998) History of the electric automobile: hybrid electric vehicles. SAE Int. ISBN 978-0-7680-0125-9
2. Eriksson R (2013) Experience from 40 years with Volvo cars efficiency and electrification. FFI Elektromobility, Göteborg
3. Turner J, Blake D, Moore J, Burke P et al (2010) The lotus range extender engine. SAE Int J Engines 3(2):318–351. doi:10.4271/2010-01-2208
4. Mattarelli E, Rinaldini C, Cantore G, Baldini P (2012) 2-Stroke externally scavenged engines for range extender applications. SAE technical paper 2012-01-1022. doi:10.4271/2012-01-1022
5. Ebner A, Winkler F, Abart M, Luz R et al (2010) Study of possible range extender concepts with respect to future emission limits. SAE technical paper 2010-32-0129. doi:10.4271/2010-32-0129
6. Bacher C, Hohenberg G (2011) Electric vehicle with combustion engine assist (EV-CEA). SAE technical paper 2011-24-0069. doi:10.4271/2011-24-0069
7. Bassett M, Thatcher I, Bisordi A, Hall, J et al (2011) Design of a dedicated range extender engine. SAE technical paper 2011-01-0862. doi:10.4271/2011-01-0862
8. Matthé R, Eberle U (2014) The Voltec system—energy storage and electric propulsion, chapter 8 in lithium-ion batteries: advances and applications. Elsevier, New York
9. Bassett M, Fraser N, Brooks T, Taylor G et al (2010) A study of fuel converter requirements for an extended-range electric vehicle. SAE Int J Engines 3(1):631–654. doi:10.4271/2010-01-0832
10. Rust A, Graf B (2010) NVH of electric vehicles with range extender. SAE Int J Passeng Cars Mech Syst 3(1):860–867. doi:10.4271/2010-01-1404

11. Govindswamy K, Tomazic D, Genender P, Schuermann G (2013) The NVH behavior of internal combustion engines used in range extended electric vehicles. SAE technical paper 2013-01-2002. doi:10.4271/2013-01-2002
12. Rabofsky A, Walther C, Rothdeutsch J, Hanfstingl R et al (2010) Electrical and hybrid drive lines: new demands for N&V integration. SAE technical paper 2010-01-1403. doi:10.4271/2010-01-1403
13. Törmänen M (2013) How to make multi-disciplinary optimization part of the product development process. Product innovation congress, Berlin, 19 Feb 2013
14. Engman J (2013) HyRange—hybridbil med en range extender i serie/parallell installation. Energirelaterad fordonsforskning 2013, Örebro, 10 April 2013
15. Arvidsson R (2014) Optimore, modular electric range extender vehicle. Optimore-GA04 meeting, Untergruppenbach, 17 March 2014

Costs and Willingness-to-Pay for Electric Vehicles

R. Kochhan and M. Hörner

Abstract Electric vehicles are hardly competitive against conventional vehicles with combustion engine today. A reason is the high cost which do not seem to match the value consumers attribute to using electric vehicles over conventional ones. In order to target the differences that exist between the costs of the electric cars and the users' willingness-to-pay (WTP), a parameter-based model has been set up. Using data from Singapore, we calculate the costs of a mid-range electric car and the users' WTP for it and confirm the assumption that the WTP is significantly lower than the costs. This difference is influenced by various technical, economic and regulatory parameters which can potentially be targeted to raise the WTP and reduce the costs. For the case of Singapore, modifying regulatory parameters like tax reductions for electric vehicles seem most suitable at today's vehicle costs, as vehicle taxes make up a large percentage of the purchase price. Modifying technical vehicle specifications like the battery capacity or the costs for them within a reasonable range does not yet equalise the costs and WTP.

1 Introduction

Compared to conventional vehicles with internal combustion engine, electric vehicles have the potential to locally reduce greenhouse gas and noise emissions. Especially in urban areas, these advantages can help to improve the quality of life, for example regarding the air quality issues of some Chinese cities today [7]. For Singapore, also the emissions over the entire life cycle of an electric vehicle are lower than of a comparable gasoline one for an accumulated mileage of more than 100,000 km. This includes the emissions of the vehicle and battery production, of the electricity production as well as of the fuel production and combustion [22].

R. Kochhan (✉) · M. Hörner
TUM CREATE Ltd., 1 CREATE Way, #10-02 CREATE Tower, Singapore 138602, Singapore
e-mail: robert.kochhan@tum-create.edu.sg

I. Denbratt et al. (eds.), *Sustainable Automotive Technologies 2014*,
Lecture Notes in Mobility, DOI 10.1007/978-3-319-17999-5_2

However, zero-emission battery electric vehicles (BEVs) do not sell well today. An important reason is that they are relatively expensive to buy, although the total cost for buying, using and reselling them may be lower than the one of conventional vehicles due to low electricity costs in comparison to fuel costs per kilometre. Another reason is the limited flexibility of a BEV as its range is typically significantly lower than the one of comparable gasoline or diesel cars, and the charging time to "refill" the battery is usually in the range of hours depending on the charging power [28]. This makes BEVs only usable for short distances and cases where enough time for charging can be guaranteed. In addition, compared to the network of petrol stations, the charging infrastructure network is not yet well developed which also reduces the flexibility of the BEV users.

Despite the technical inconveniences of BEVs or the missing infrastructure, costs are often decisive. In particular, this leads to the question about the amount of the total costs of BEVs, and under which circumstances this amount can be lower than the actual willingness-to-pay (WTP) of people for those environmentally friendly cars.

Therefore, a model has been set up which allows comparison of the costs of BEVs and the customers' WTP for them. The purpose is to quantify the "gap" between the costs and the WTP and to find possibilities to reduce it. The comparison is based on a range of parameters which have an effect on the WTP of potential customers, on the total cost of ownership (TCO) of the electric vehicles or on both of them. Varying these parameters can affect the gap.

The objective is to identify promising parameters which allow to minimise the gap effectively. For this analysis, we use TCO and WTP data from Singapore in order to show the potential of the approach. The results can be a basis for subsequent work which should aim at a refinement of the approach and at the feasibility of measures to change the identified parameters in order to reduce the gap.

2 Total Cost of Ownership of Electric Vehicles

There is no general definition of the TCO [6]. It has to be applied according to the use case. For this analysis, the TCO is split into three main components [3, 19]: the acquisition costs, the operating costs and the end-of-life value of the car. For this study, a former TCO model presented in [10] has been modified and integrated into a combined TCO-WTP model.

As this study was conducted in Singapore, the acquisition costs of cars include a range of parameters which are specific for this city. Most of them are taxes which are imposed on the open market value of a vehicle and which make cars in Singapore relatively expensive. An important tax is the certificate of entitlement (COE) which a potential car buyer has to obtain in a bidding system [13–15]. The purpose of the COE is to limit the traffic volume on the streets in Singapore in order to limit traffic congestions [10].

Table 1 Key specifications and costs of the analysed vehicles [1, 2, 4, 5, 11, 12, 14–17, 20, 21, 27–29]

General settings and assumptions		
Year of purchase	(Beginning of) 2013	
Discount rate	3 %	
Annual mileage	18,200 km	
Usage period	7 years	
Main vehicle specifications		
	Volkswagen Golf	Volkswagen e-Golf
Fuel/electricity consumption	7.5 L/100 km	16.5 kWh/100 km
Engine/motor power	103 kW	85 kW
Battery capacity		24.2 kWh
Main TCO components in USD (rounded values)		
	Volkswagen Golf	Volkswagen e-Golf
Purchase price	130,276	157,195
Thereof OMV	23,321	33,448
Thereof COE	59,699	64,166
Thereof ARF	26,278	40,433
Thereof CEVS rebate	−11,989	−15,986
Operating costs	49,129	38,540
Thereof fuel/electricity costs	14,521	4,134
End-of-life value	−30,749	−43,333

All values are taken from the given sources or are calculations and estimations based on them

OMV Open market value, *COE* certificate of entitlement, *ARF* additional registration fee, *CEVS* carbon emissions-based vehicle scheme

The operating costs include insurance premiums, maintenance costs, cost for fuel or electricity, an annual vehicle tax as well as average parking costs. In addition, some other cost components may increase the operating costs, for example using toll roads.

The end-of-life or resale value is estimated to be relatively high because the high acquisition costs lead to an increased resale value after 7 years of usage, including tax returns. For example, as the COE is valid for 10 years, it increases the end-of-life value of the vehicle if it is sold or deregistered after only 7 years [15]. The end-of-life value is given as negative costs, assuming that the user can earn money by selling the car at the end of the 7-year usage period.

A detailed breakdown of the different components of the acquisition costs is shown in Table 1.

Emissions are important for the calculation of the vehicle cost as well. In the case of Singapore, the tax rebates (CEVS) are granted depending on the vehicles' tank-to-wheel emissions encouraging car buyers to make their decision for more environmentally friendly cars. Emissionless electric vehicles are granted the maximum rebate on the purchase price of almost 16,000 USD [15] (Table 1).

3 Willingness-to-Pay for Electric Vehicles

The willingness-to-pay (WTP) can be defined as the maximum price which a consumer would be willing to spend for a product (e.g. [24]). In this case, the WTP is the amount of money a potential customer is willing to spend for buying a BEV.

In order to estimate this WTP, a consumer survey in Singapore was conducted. Participants were asked to rank a range of mobility-relevant attributes according to how important they are to them when travelling around in Singapore. The attributes which were more important to the respondents were then used in a second web-based survey in which 914 Singaporeans were asked about their WTP for vehicle-related parameters which are characteristic of the particular mobility attributes [10]. Table 2 shows the mobility attributes and vehicle parameters selected for this study.

In order to quantify the WTP, the method of a choice-based conjoint analysis was used in the survey. This approach allows to estimate utility values for different levels of the parameters listed in Table 2. Using the purchase price (premium) as one of the parameters, monetary WTP values were calculated for each parameter level. This survey design does not allow to estimate a large number of WTP data points per parameter. Hence, the WTP curves were extrapolated following the law of diminishing marginal utility using a simple logarithmic function [9, 18].

Figure 1 exemplarily shows the ΔWTP for the mobility attribute "flexibility", described by the parameter "driving range". The ΔWTP is the price premium a consumer is willing to pay on top of the purchase price of a comparable

Table 2 Mobility attributes and parameters

Mobility attributes	BEV parameters	Parameter unit
Time savings	Time to fully charge the battery	hours
Reliability and certainty	Distance to next available charging station	km
Convenience	Availability of home charging	Yes/no
Cleanliness	Amount of inner-urban pollution reduced	% Relative to conventional vehicle
Flexibility	Maximum driving range on a single charge	km
Costs	Reduction of operating costs	% Relative to conventional vehicle
Costs	Purchase price premium	Fixed monetary amount relative to conventional vehicle

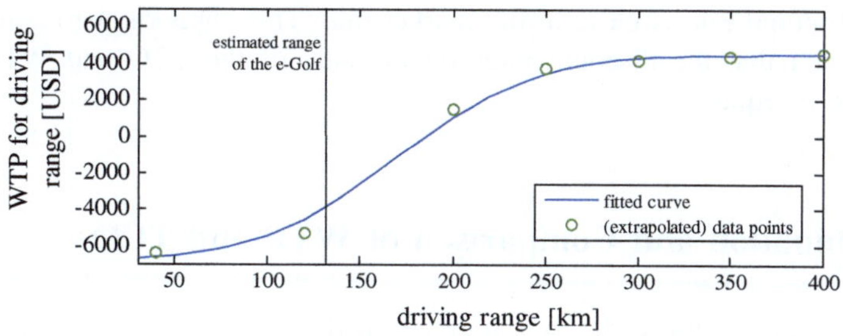

Fig. 1 Example of a ΔWTP curve for driving range

conventional vehicle (Volkswagen Golf). Ideally, as described by [8] for the US, the WTP for driving range is negative for ranges which are less than most of today's conventional cars. According to the data collected in Singapore, consumers already value ranges of above 200 km positively. This difference is certainly related to the different survey design, but is certainly also influenced by the relatively short distances within the limited area of Singapore, so that lower ranges are considered sufficient for the daily usage of electric cars.

Based on the ΔWTP values for the different parameters, the total WTP for an electric vehicle WTP_{BEV} is calculated by summing up all ΔWTP values for parameter configurations of this BEV and the purchase price of the reference conventional vehicle:

$$WTP_{BEV} = \Sigma\, \Delta WTP_{BEV,i} + C_{purchase,ref.vehicle}$$

with $\Delta WTP_{BEV,i}$ being the purchase price premiums regarding the parameters i of the electric vehicle (Table 2) and $C_{purchase,\ ref.\ vehicle}$ the acquisition costs (purchase price) of the conventional reference vehicle. In theory, WTP_{BEV} of a potential customer for an electric car has to be equal or higher than its acquisition costs so that the car can be sold.

In this survey design, only the acquisition costs can be easily compared to the WTP for a BEV, as the WTP is the total amount a person would be willing to spend to buy an electric vehicle, and not both to buy and use it for several years. This fact seems to make the comparison incomplete because only a part of the TCO is taken into consideration, whereas the customer would have to come up for all costs when using the car, and not only for the acquisition costs. However, the comparison of the WTP and acquisition costs already takes into account the operating costs of a car which are one of the parameters used in the WTP survey (Table 2). Hence, a modification of the statement about the operating costs of a vehicle configuration in the survey would also affect the respondents' answers. Higher operating costs are generally perceived as a disadvantage.

This implicates that the consumers take into account an implicit discount rate when weighing the low future operating costs against the high acquisition costs. Verboven [25] shows for similar trade-offs between different vehicle types that this implicit discount rate is close to the market rate. This supports the assumption in our approach that the discount rates on the side of the TCO and WTP can be considered as equal.

4 Combination and Comparison of WTP and TCO

Both the TCO and the WTP data are combined in a parameter-based model. The goal is to compare the values of the acquisition costs and the WTP for electric and conventional vehicles and to specify the difference between both values ("gap") using parameters which describe the TCO, the WTP or both of them.

The parameters can be classified in different categories. Vehicle-related or technical parameters include the electric motor or engine power, the battery capacity of the BEV or the fuel and electricity consumptions. Many cost values like battery costs, the motor/engine production costs as well as margins are attributed to the vehicle manufacturer. Taxes, incentives or registration fees which have a significant impact on the TCO of vehicles in Singapore are summarised in a third category.

Some of these parameters influence either the TCO or the WTP. However, the parameters which have an impact on both are considered most interesting. For example, a lower battery capacity decreases the purchase price of a vehicle, but also changes the WTP as the range of electric vehicles is an important aspect for potential customers.

Based on the estimated values of the TCO and the WTP, a cost "gap" between both values is calculated as follows:

$$C_{gap} = C_{BEV} - WTP_{BEV}$$

with C_{BEV} being the acquisition costs of the electric vehicle and C_{gap} the monetary difference between the consumers WTP_{BEV} for the electric vehicle and its purchase price.

Ideally, WTP_{BEV} surpasses C_{BEV} resulting in a negative C_{gap}. However, for the case of Singapore, we confirmed the assumption that this is generally not the case today. We analysed the example of a Volkswagen Golf, which is also available as an electric version. The costs of this car are shown in Table 1. Figure 2 displays the acquisition costs and the WTP for both versions in Singapore. The WTP for the conventional Golf is assumed to be as high as the acquisition costs because it has already been sold in Singapore at this price (in reality, the WTP can be equal or higher than the costs), whereas the WTP for the e-Golf is estimated based on the survey data.

According to the data, the WTP for the Volkswagen e-Golf is approx. 18,500 USD lower than the purchase price of the conventional Volkswagen Golf, and significantly lower (approx. 45,500 USD) than the estimated purchase price of an e-Golf in Singapore. In general, this relationship seems reasonable as an e-Golf can

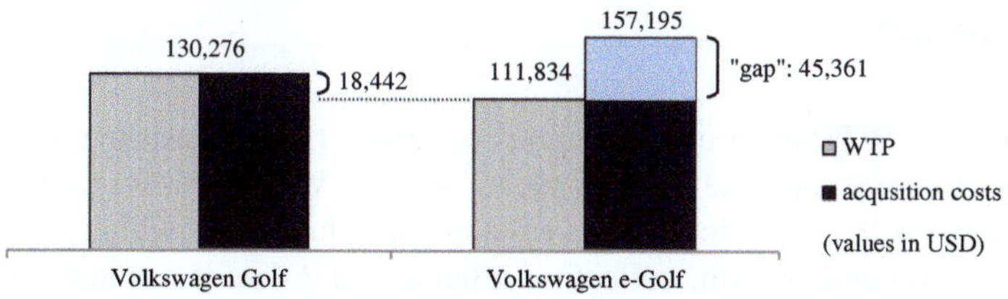

Fig. 2 Estimated WTP for and acquisition costs of an electric and conventional Volkswagen Golf in Singapore

certainly be regarded as less convenient as a conventional Golf considering the currently available infrastructure, the short range of about 130 km (Fig. 1) and the long charging time. The absolute height of these values may be questionable, but regarding the high vehicle prices in Singapore including vehicle taxes which are a lot higher than the gap of 45,500 USD, these numbers are within an acceptable range.

Consequently, the question arises which parameters are mainly responsible for the gap, and which possibilities exist to change them so that C_{gap} decreases. Therefore, the three categories described above were distinguished: technical vehicle-related aspects, parameters which have to be attributed to the vehicle manufacturer, and governmental or regulatory aspects. Changing the electric vehicle specifications like battery capacity or electricity consumption in a way that costs are reduced and the WTP of potential customers increases fall under the first category. In this case, the gap could be reduced by trying to modify the BEV technically in order to lower its costs and increase its attractiveness for potential customers at the same time, resulting in a higher WTP. Parameters of the second and the third category focus directly on the costs rather than on technical vehicle aspects. On the side of the manufacturer, costs of the vehicle or vehicle components can be reduced for example by technological progress or high-volume production.

In Singapore, regulatory levers seem to be most promising regarding the enormous vehicle taxation. Exempting BEVs from only one of the major vehicle taxes ARF or COE can make the vehicle purchase price of a Volkswagen e-Golf drop by at least 40,000 USD which would almost eliminate the gap. In contrast, the sensitivity of the gap to changes of technical vehicle parameters is comparatively low.

However, targeting vehicle taxation is not considered as a long term strategy, but rather as an incentive in order to motivate users to get in touch with the new technology. In addition, as the price of COE is determined in a bidding system, future COE prices are not directly controllable by the government which makes it more difficult to prioritise electric vehicles through the COE system. This uncertainty weakens the effect of those regulations or at least narrows the possibilities of governmental interventions to other taxes like the ARF. This aspect should be addressed in future studies.

5 Conclusion

A consumer's "WTP is an unobservable construct" [26]. Measuring it can only be an attempt to come as close as possible to the true WTP [26]. Regarding this fact and the situation that electric vehicles are a new product today which potential users are hardy very familiar with, collecting reliable WTP data is a major challenge of this approach. In addition, the TCO can also only be estimated using plausible assumptions about vehicle usage and future price developments. Hence, the approach described in this paper is considered to be suitable to get a rough overview of the suitability of electric vehicles for selected markets or users. An exact comparison of costs of cars and WTP for them will not be possible. However, validating the stated WTP data by additional interviews, experiments or observed WTP data will be useful.

Regarding the model, the parameters which are used to estimate and compare the costs and WTP of electric cars have to be synchronised in order to be able to simulate the effects that certain technical and regulatory changes have on both costs of the electric vehicles and the consumers' WTP for it. This can help to understand a BEV in its environment and to tailor the measures to introduce BEVs in selected markets. The example of Singapore may be a special case regarding the vehicle costs in this city state, but it still seems to be a promising market for electric vehicles where short distances and inner-urban traffic conditions prevail, which are generally more favourable for BEVs than for conventional vehicles.

At long term, it is certainly economically more viable to focus on measures which help to increase the attractiveness of BEVs for consumers on a technical or infrastructure level than trying to push the costs below a certain level by giving for example tax incentives. Advertising lower operating costs of BEVs may also fail. According to [23], people usually make their opinion about the acquisition costs of private vehicles in a different way than estimating future expenses and payback periods as they do not have the basic building blocks to make a calculated decision. However, at short term, incentives can be helpful to let more people get in touch with the new technology.

Acknowledgments The work presented was financially supported by the Singapore National Research Foundation under its programme "Campus for Research Excellence and Technological Enterprise" (CREATE).

References

1. AECOM Australia (2009) Economic viability of electric vehicles. Report prepared for the Department of Environment and Climate Change, Sydney, Sep 2009
2. Baum H, Dobberstein J, Schuler B (2011) Nutzen-Kosten-Analyse der Elektromobilität. Institute of Transportation Sciences, University of Cologne, Feb 2011
3. Contestabile M, Offer GJ, Slade R, Jaeger F et al (2011) Battery electric vehicles, hydrogen fuel cells and biofuels. Which will be the winner? Energy Environ Sci 4:3754–3772. doi:10.1039/c1ee01804c

4. Douglas C, Stuart A, Dolman M (2011) Influences on the low carbon car market from 2020–2030, final report for low carbon vehicle partnership. Element Energy Ltd., Cambridge
5. Electrification Coalition (2013) EV market outlook—state of the plug-in electric vehicle market. Report, Washington DC
6. Ellram LM, Siferd SP (1998) Total cost of ownership: a key concept in strategic cost management decisions. J Bus Logist 19(1):55–84
7. Han X, Naeher LP (2006) A review of traffic-related air pollution exposure assessment studies in the developing world. Environ Int 32:106–120. doi:10.1016/j.envint.2005.05.020
8. Hidrue MK, Parsons GR, Kempton W, Gardner MP (2011) Willingness to pay for electric vehicles and their attributes. Resour Energy Econ 33:686–705. doi:10.1016/j.reseneeco.2011.02.002
9. Johnson TC (2007) Utility functions. C2922 economics, Heriot Watt University, Edinburgh
10. Kochhan R, Lim J, Knackfuß S, Gleyzes D et al (2013) Total cost of ownership and willingness-to-pay for private mobility in Singapore. In: Proceedings of the international conference of future automotive technologies, Ingolstadt
11. Kochhan R, Fuchs S, Reuter B, Burda P et al (2014) An overview of costs for vehicle components, fuels and greenhouse gas emissions
12. Land Transport Authority of Singapore (2013) Costs (S$) for cars registered in 2013. Data sheets from Jan 2013 through Dec 2013
13. Land Transport Authority of Singapore (2014) Certificate of entitlement (COE), Singapore. http://www.lta.gov.sg/content/ltaweb/en/roads-and-motoring/owning-a-vehicle/vehicle-quota-system/certificate-of-entitlement-coe.html. Accessed 15 May 2014
14. Land Transport Authority of Singapore (2014) COE bidding results 2013, Singapore
15. Land Transport Authority of Singapore (2014) Tax structure for cars. http://www.lta.gov.sg/content/ltaweb/en/roads-and-motoring/owning-a-vehicle/costs-of-owning-a-vehicle/tax-structure-for-cars.html. Accessed 05 May 2014
16. Land Transport Authority of Singapore (2014) Singapore land transport statistics in brief 2013, Singapore
17. Moore RJ (2011) Parking rate survey—parking rates broadly up—worldwide. Colliers International, Seattle
18. Norstad J (1999) An introduction to utility theory. Northwestern University, USA
19. NPE (2011) Zweiter Bericht der Nationalen Plattform Elektromobilität. Anhang, Abb. A05, Berlin
20. OANDA (2014) Historical exchange rates. http://www.oanda.com/currency/historical-rates. Accessed 15 May 2014
21. Propfe B, Redelbach M, Santini DJ, Friedrich H (2012) Cost analysis of plug-in hybrid electric vehicles including maintenance and repair costs and resale values. In: Presented at the EVS26 international battery, hybrid and fuel cell electric vehicle symposium, Los Angeles
22. Reuter B, Schönsteiner K, Wagner M, Gleyzes D et al (2014) Life cycle greenhouse gas analysis for automotive applications—a case study for taxis in Singapore. Int J Smart Grid Clean Energy 3(2):127–134. doi:10.12720/sgce.3.2.127-134
23. Turrentine T, Kurani K (2007) Car buyers and fuel economy? Energy Policy 35:1213–1223. doi:10.1016/j.enpol.2006.03.005
24. Varian HR (1992) Microeconomic analysis. Norton, New York. ISBN 0-393-95735-7
25. Verboven F (1999) Implicit interest rates in consumer durables purchasing decisions—evidence from automobiles. University of Antwerp
26. Voelckner F (2006) An empirical comparison of methods for measuring consumers' willingness to pay. Market Lett 17:137–149. doi:10.1007/s11002-006-5147-x
27. Volkswagen (2014) Der Golf. Das Auto. Technik und Preise
28. Volkswagen (2014) Der neue e-Golf. Technik und Preise
29. Wietschel M, Bünger U, Weindorf W (2010) Vergleich von Strom und Wasserstoff als CO_2-freie Energieträger. Fraunhofer ISI, Karlsruhe

Part II
Vehicle Dynamics

Better Road Design Using Clothoids

Hormoz Marzbani, Reza N. Jazar and M. Fard

Abstract According to the mobile robot researchers and experts the best and smoothest transition curve to be used as a section of the path is the Eulers Spiral also known as Clothoid. During 19th century Arthur Talbot derived the equation of Clothoids to use as an easement curve for the purpose of avoiding the shock and disagreeable lurch of trains, due to instant change of direction. The Euler Spiral is a curve whose degree-of-curve increases directly with the distance along the curve from the start point of the spiral. This will provide a linear change in the steering angle required by the driver to go through the turn. In other word for a car which is travelling on a Clothoid transition curve of the road there will be no need for sudden changes in the steering angle of the wheels. The angle required starts from zero and increases to a maximum value and back to zero linearly. This provides a very comfortable ride for the passengers of the vehicle. The use of these curves for road design have been studied here, and a design chart have been proposed to be used for finding the best suitable transition curve for different applications.

Keywords Tire kinematics · Tireprint · Contact patch · Rolling tire · Radial deflection

H. Marzbani (✉) · R.N. Jazar · M. Fard
School of Aerospace, Mechanical and Manufacturing Engineering, RMIT University, Melbourne, Australia
e-mail: hormoz.marzbani@rmit.edu.au

R.N. Jazar
e-mail: reza.jazar@rmit.edu.au

M. Fard
e-mail: mohammad.fard@rmit.edu.au

I. Denbratt et al. (eds.), *Sustainable Automotive Technologies 2014*,
Lecture Notes in Mobility, DOI 10.1007/978-3-319-17999-5_3

1 Introduction

Clothoid or "Cornu Spiral", Also known as "Euler's Spiral" have been introduced as the answer to finding the best transition curve. The transition curve as a section of a road is where most of the researchers referred to this type of planar curve as the answer for getting the smoothest ride possible. Primary function of a transition curve or easement curve is to accomplish gradual transition from a straight line to a circular curve, so that curvature changes from zero to a finite given value. To call a spiral between a straight and a curve as a valid transition curve, it has to satisfy the following conditions:

- One end of the spiral should be tangential to the straight
- The other end should be tangential to the curve
- Spiral's curvature at the intersection point with the circular arc should be equal to arc curvature
- The rate of change of curvature along the transition should be same as that of the increase of cant
- Its length should be such that full cant is attained at the beginning of circular cant

The research on trajectory design for mobile robots' path of motion, can be used to help designing better roads.

Several trajectory design methods have been presented using simple curve elements, such as straight lines, circles, clothoids, and their combinations. It is important that a trajectory forms a smooth curve because the mass of a robot—or a vehicle in the case of road design—must move along the designed curve. If the curve is not smooth enough, undesirable effects occur such as side slip or deviation from the course at points where the curvature along the curve is not continuous. The history of trajectory design can be viewed as attempts to avoid this inconvenience by using smoother curves [1].

Kanayama [2] believes that including Clothoids in designing a curve, makes curvature control feasible.

He used a clothoid pair to make curves with zero curvature at their junctures with line segments in order to produce continuous drive-velocity functions for a differential drive cart. Clothoid pairs have the advantage of providing the minimum-length curves for a given limit on jerk; they have the disadvantage, however, that the (x, y) coordinates of the curve have no closed-form expressions, but must be derived by integrating along the path length, s, with the curvature $c(s)$ linearly increasing from zero during the first half of the turn, then linearly decreasing to zero during the second half, in such a way the end-point of the curve precisely matches the position and slope constraints of the succeeding line segment.

It is known today, based on the experience of the past 50 years, that the alignment of modern highways has to be consistent and efficient from a driving dynamic and convincing from a driving psychological point of view. In this connection the Clothoid, as a transition curve, offers good solutions.

The clothoid satisfies esthetical solutions and by being flexible, enables a good adaptation to the topography and existing local constraints. The clothoid guarantees for motor vehicles an economically efficient ride, and saves, through it's appropriate insertion into the local environment, considerable construction costs [3].

2 History and Definition of Clothoids

Cornu Spirals were found for the first time, long before being used for path generation for different applications.

Euler Spiral, is beautiful and useful curve known by several other names, including "Clothoid" and "Cornu Spiral." The underlying mathematical equation is also most commonly known as the Fresnel Integral. The profusion of names reflects the fact that the curve has been discovered several different times, each for a completely different application: first, as a particular problem in the theory of elastic springs; second, as a graphical computation technique for light diffraction patterns; and third, as a railway transition spiral which is the closest application to the interest of the topic of this thesis.

The Euler spiral is defined as the curve in which the curvature increases linearly with arc length. Equations 2 and 3 show the general form of the parametric equation of this curve, which are known as the Fresnel integrals.

Considering curvature as a signed quantity, it forms a double spiral with odd symmetry, a single inflection point at the center, as shown in Fig. 1. The first appearance of the Euler spiral is as a problem of elasticity, posed by James Bernolli in the same 1694 publication as his solution to a related problem, that of elastica.

The elastica is the shape defined by an initially straight band of thin elastic material (such as spring metal) when placed under load at its endpoints. The Euler spiral can be defined as something of the inverse problem; the shape of a pre-curved spring, so that when placed under load at one endpoint, it assumes a straight line, as shown in Fig. 2.

Fig. 1 Euler spiral

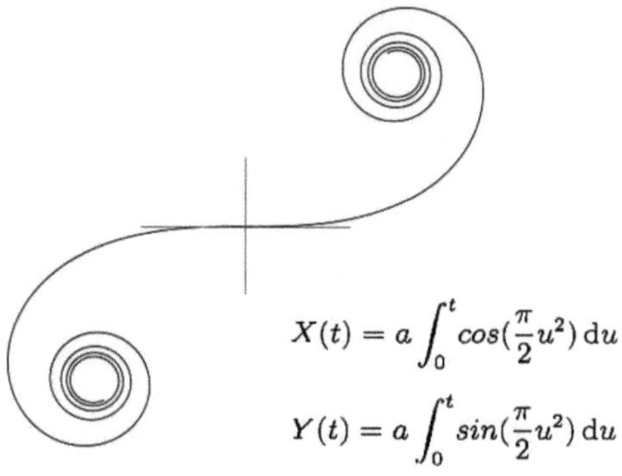

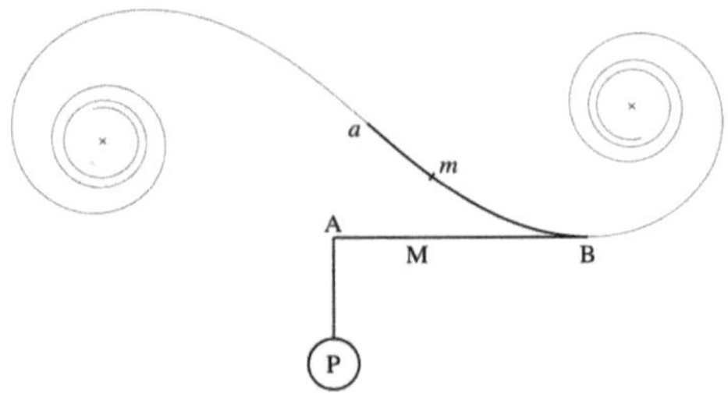

Fig. 2 Reconstruction of Euler's figure, with complete spiral superimposed

In 1744, Euler rephrases the problem as: What shape must the lamina *amB*—see Fig. 2—take so that it is flattened into an exactly straight line when the free end is pulled down by weight p?

The answer derived by Euler appeals to the simple theory of moments: The moment at any point M along the lamina is the force p times the distance s from A to M. The curvature of the curve resulting from the original shape stressed by force p is equal to the original curvature plus the moment P_s divided by the lamina's stiffness Ek^2. Since this resulting curve must be a straight line with curvature zero, the solution for the curvature of the original curve is $\kappa = -Ps(Ek^2)^{-1}$. Euler flips the sign for the curvature and groups all the force and elasticity constants into one constant a for convenience, yielding $1/r = \kappa = s/aa$. From this intrinsic equation, Euler derives the curve's quadrature (x and y as a function of the arc length parameter s), giving equations for the modern expression of the curve:

$$x = \int \sin \frac{ss}{2aa}\, ds \qquad \text{and} \qquad y = \int \sin \frac{ss}{2aa}\, ds \tag{1}$$

It took Euler about 38 years to solve the problem of the integral's limits. In his 1781 "On the values of integrals extended from the variable term $x = 0$ up to $x = \infty$", he finally gave the solution, which he had "recently found by a happy chance and in an exceedingly peculiar manner", of $x = y = \frac{a}{\sqrt{2}}\sqrt{\frac{\pi}{2}}$. Around 1818, Augustine Fresnel considered a problem of light diffracting through a slit, and independently derived integrals equivalent to those defining the Euler spiral. At the time, he seemed to be unaware of the fact that Euler (and Bernoulli) had already considered these integrals, or that they were related to a problem of elastic springs. Later, this correspondence was recognized, as well as the fact that the curves could be used as graphical computational method for diffraction patterns.

The third completely independent discovery of the Euler spiral is in the context of designing railway tracks to provide a smooth riding experience. Over the course of the 19th century, the need for a track shape with gradually varying curvature became clear. Arthur Talbot was among the first to approach the problem mathematically, and derived exactly the same integrals as Bernoulli and Fresnel before him. His introduction to "The Railway Transition Spiral" [4] describes the problem and his solution articulately: A transition curve, or easement curve, as it is sometimes called, is a curve of varying radius used to connect circular curves with tangents for the purpose of avoiding the shock and disagreeable lurch of trains, due to the instant change of direction and also to the sudden change from level to inclined track. The primary object of the transition curve, then, is to affect smooth riding when the train is entering or leaving a curve. The generally accepted requirement for a proper transition curve is that the degree of curve shall increase gradually and uniformly from the point of tangent until the degree of the main curve is reached. In addition to this, an acceptable transition curve must be so simple that the field work may be easily and rapidly done, and should be so flexible that it may be adjusted to meet the varied requirements of problems in location and construction. "The Transition Spiral is a curve whose degree-of-curve increases directly as the distance along the curve from the point of spiral."

3 Clothoids in Road Design

The main area of interest in the existing study is the use of these type of planar curves in designing roads. Roads are made by continuously connecting straight and circular paths by proper transition turning sections. The transition curve in represented by the Euler spiral curve applied to effect the transition between two circular curves or between a circular curve and a tangent. The transition curve will:

- Provide a linear gradual increase or decrease of the centrifugal acceleration for the transition from one design element to the other when passing through curves
- Serve as a transition section for a convenient desirable arrangement for the superelevation runoff
- Make possible through the gradual change of the curvature a consistent alignment and through this a consistent operating speed
- Create a satisfactory optical appearance of the alignment

As mentioned earlier the *clothoid spiral* is the best smooth transition connecting curve in road design which is expressed by parametric equations called *Fresnel Integral*:

$$X(t) = a \int_0^t \cos(\frac{\pi}{2} u^2)\,\mathrm{d}u \tag{2}$$

$$Y(t) = a \int_0^t \sin(\frac{\pi}{2} u^2)\,\mathrm{d}u \tag{3}$$

The curvature of the clothoid curve varies linearly with arc length and this linearity makes clothoid the smoothest driving transition curve. The scaling parameter a is only a magnification factor that shrinks or magnifies the curve. The range of t determines the variation of curvature within the clothoid, as well as the initial and final tangent angles of the clothoid curve. The *arc length, s*, of a clothoid for a given value of t is

$$s = at \tag{4}$$

If the variable t indicates time then, a would be the speed of motion along the path. The *curvature k* and *raius of curvature R* of a clothoid at a given t is

$$k = \frac{\pi t}{a} \tag{5}$$

$$R = \frac{1}{k} = \frac{a}{\pi t} \tag{6}$$

The tangent angle θ of a clothoid with a given value t is

$$\theta = \frac{\pi}{2} t^2 \tag{7}$$

Having a road with linearly increasing curvature is equivalent to entering the path with a steering wheel at the neutral position and turning the steering wheel with a constant angular velocity. This is a desirable and natural driving action. One of the most common scenarios, for when to use clothoids in road design is for connecting a straight road to a circle. The following is an example, for the mentioned scenario.

3.1 Connecting a Straight Road to a Circle Using a Clothoid

Assume that we need to define a clothoid road to connect a straight line to a circle. In other words we need to find a clothoid which has a zero curvature to start with, and meet up with a circular curve at the end which means it need to have the same curvature as the curve at the end point. For simplifying we can assume the straight part of the road is the X-axis, and then if needed transfer or rotate it to the required place and orientation. The circle used in this example have a radius of $R = 100$ m at the center point $C(62.811, 106.658)$.

$$(X - 62.811)^2 + (Y - 106.658)^2 = 100^2 \tag{8}$$

As mentioned earlier the transition road must begin with $k = 0$ on the X-axis and touch the circle at a point when its curvature is $k = 1/R = 1/100$. Using

$$k = \frac{\pi s}{a^2} = \frac{\pi t}{a} \tag{9}$$

one can writhe the parametric equation of the clothoid to be used as the following:

$$X(t) = a\int_0^{\frac{ka}{\pi}} \cos(\frac{\pi}{2}u^2)\,\mathrm{d}u \tag{10}$$

$$Y(t) = a\int_0^{\frac{ka}{\pi}} \sin(\frac{\pi}{2}u^2)\,\mathrm{d}u \tag{11}$$

Knowing the curvature at the end point of the clothoid, $k = 0.01$, we can find the coordinates of the end point as a function of a.

$$X(t) = a\int_0^{\frac{0.01a}{\pi}} \cos(\frac{\pi}{2}u^2)\,\mathrm{d}u, \quad Y(t) = a\int_0^{\frac{0.01a}{\pi}} \sin(\frac{\pi}{2}u^2)\,\mathrm{d}u \tag{12}$$

In the next step, we need to find the magnifying factor, a, such that the clothoid meets the circle with the same slope. The slope of the circle at (X, Y) is

$$Y' = \tan\theta = -\frac{X - 62.811}{Y - 106.658} \tag{13}$$

and the slope angle of the clothoid at the end point, where it meets the circle is

$$\theta = \frac{\pi}{2}t^2 = \frac{1}{2\pi}a^2k^2 = 1.5915 \times 10^{-5}a^2 \tag{14}$$

By equating the slope angles of the circle with the end point of the clothoid we'll end up with an equation with respect to a.

$$\arctan\frac{-(X - 62.811)}{Y - 106.658} = 1.5915 \times 10^{-5}a^2 \tag{15}$$

Using Eqs. 8, 12, and 15, we get and equation which a can be driven from.

$$\begin{aligned}
&\arctan\frac{-(X - 62.811)}{Y - 106.658} - 1.5915 \times 10^{-5}a^2 \\
&= \arctan\frac{-(X - 62.811)}{\sqrt{100^2 - (X - 62.811)^2}} - 1.5915 \times 10^{-5}a^2 \\
&= \arctan\frac{-(a\int_0^{\frac{0.01a}{\pi}} \cos(\frac{\pi}{2}u^2)\,\mathrm{d}u - 62.811)}{\sqrt{100^2 - (a\int_0^{\frac{0.01a}{\pi}} \cos(\frac{\pi}{2}u^2\,\mathrm{d}u - 62.811)^2)}} - 1.5915 \times 10^{-5}a^2
\end{aligned} \tag{16}$$

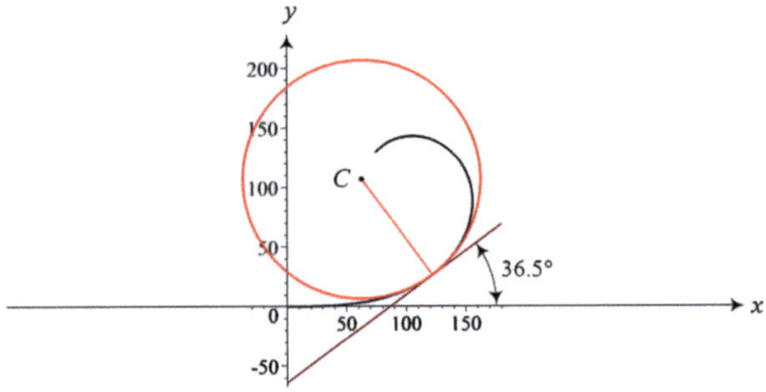

Fig. 3 The tangent line, normal line, and the tangent circle to the clothoid at the point where $k = 0.01$ for a given $a = 200$. The clothoid is plotted up to $k = 0.025$

solving Eq. 22, a will be found, which in this case:

$$a = 200 \tag{17}$$

Having a, the equations of the clothoid can be defined. As mentioned before having the curvature of the end point of the clothoid one can find the point which the clothoid and the circle intersect.

$$X_0 = 122.2596310 \qquad Y_0 = 26.24682756 \tag{18}$$

The slope of the road at the point, the tangent line to the road and the normal line to the road are

$$\theta = \frac{1}{2\pi} a^2 k^2 = 0.6366197722\,\text{rad} = 369.475° \tag{19}$$

$$Y = -64.14007833 + 0.7393029502X \tag{20}$$

$$Y = 191.6183183 - 1.352625469X \tag{21}$$

Figure 3 illustrates the clothoid, tangent line, normal line and the tangent circle to the clothoid at point (X_0, Y_0). Here is another example for finding the right transition curve, to connect a straight road to a circle. This time we'll try to find the appropriate transition curve to connect a straight road as *X-axis*, with zero curvature to a circle of radius $R = 80$ m at center $c(100, 100)$.

$$(X - 100)^2 + (Y - 100)^2 = 80^2 \tag{22}$$

Using

$$X(t) = a\int_0^{\frac{ka}{\pi}} \cos(\frac{\pi}{2}u^2)\,\mathrm{d}u \tag{23}$$

$$Y(t) = a\int_0^{\frac{ka}{\pi}} \sin(\frac{\pi}{2}u^2)\,\mathrm{d}u \tag{24}$$

and knowing $k = 1/R = 0.0125$ at the destination point, determines the coordinates of the end of the clothoid as function of a.

$$X(t) = a\int_0^{\frac{0.0125a}{\pi}} \cos(\frac{\pi}{2}u^2)\,\mathrm{d}u \tag{25}$$

$$Y(t) = a\int_0^{\frac{0.0125a}{\pi}} \sin(\frac{\pi}{2}u^2)\,\mathrm{d}u \tag{26}$$

The slope of the tangents to the circle 22, at a point (X, Y) is

$$Y' = \tan\theta = -\frac{X - 100}{Y - 100} \tag{27}$$

and the slope angle of the clothoid at its end point as a function of a is,

$$\theta = \frac{\pi}{2}t^2 = \frac{1}{2\pi}a^2k^2 = 2.4868 \times 10^{-5}a^2 \tag{28}$$

It is important that the clothoid and circle have the same slope angle at the end point of the clothoid, which leads us to the following equation by equating the two slope angles.

$$\arctan\frac{-(X - 100)}{Y - 100} = 2.4868 \times 10^{-5}a^2 \tag{29}$$

Equations 22 and 29 along with the equation of the clothoid, gives us as equation with respect to a. However, substituting $Y = Y(X)$ and replacing tan and arctan generates four equations to be solved for possible a. To visualize the possible solutions, let us define four error equations.

$$e = \arctan\frac{-(X - 100)}{\pm\sqrt{80^2 - (X - 100)^2} - 2.4868 \times 10^{-5}a^2} \tag{30}$$

$$e = \frac{-(X - 100)}{\pm\sqrt{80^2 - (X - 100)^2} - \tan(2.4868 \times 10^{-5}a^2)} \tag{31}$$

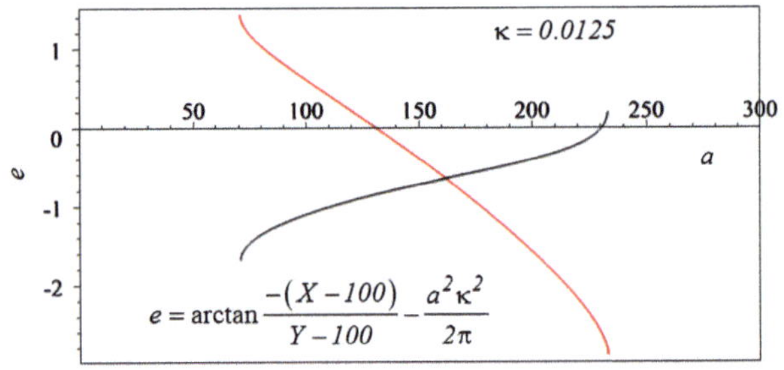

Fig. 4 Plot of $e = \arctan \frac{-(X-100)}{\pm\sqrt{80^2-(X-100)^2-2.4868\times10^{-5}a^2}}$ versus a

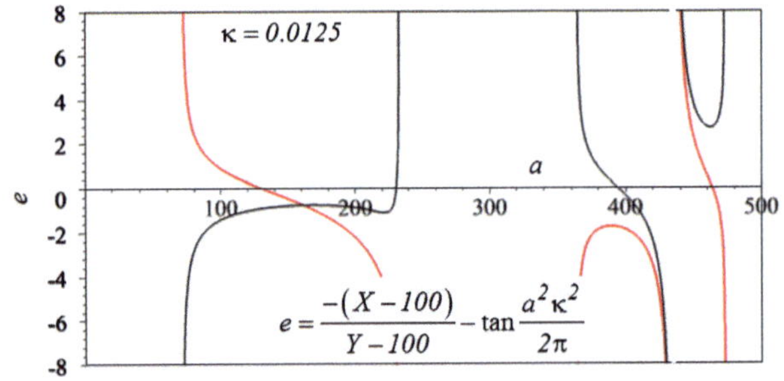

Fig. 5 Plot of $e = \frac{-(X-100)}{\pm\sqrt{80^2-(X-100)^2-\tan(2.4868\times10^{-5}a^2)}}$ versus a

Figure 4 depicts Eq. 30 and Fig. 5 shows Eq. 31. From the first one we get the solutions,

$$a = 230.7098693 \qquad a = 130.8889343 \tag{32}$$

and the second equation gives the following as the solutions,

$$a = 230.7098693 \quad a = 130.8889343 \quad a = 394.0940573 \quad a = 463.5589702 \tag{33}$$

The correct answer is $a = 230.7098693$ and Fig. 6 depicts the circle and the proper clothoid, Having a defines the clothoid equation which at $k = 0.0125$ reaches to

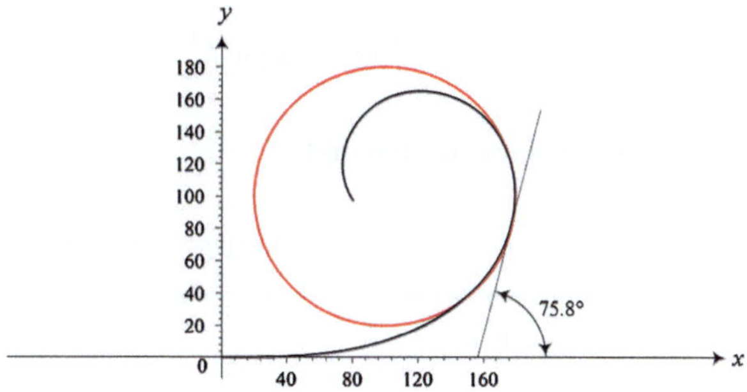

Fig. 6 The transition road starting on *X*-axis and goes to a circle of radius *R* = 80 m at center *c*(100, 100)

$$X_0 = 177.5691613 \qquad Y_0 = 82.38074640 \tag{34}$$

at angle,

$$\theta = \frac{1}{2\pi} a^2 k^2 = 1.3236\,\text{rad} \approx 75.84^\circ \tag{35}$$

There are exceptions using the method which has been introduced here, as well. It is sometimes needed that we shift the clothoid in order to meet the given circle. It is not generally possible to design a clothoid starting at the origin and meet a given circle at an arbitrary center and radius. However, it is possible to start the clothoid from other points on the *X-axis* rather than the origin to meet the given circle. Following is the equation of the given circle for which we need to find a transition clothoid curve,

$$(x - x_c)^2 + (y - y_c)^2 = R^2 \tag{36}$$

where (x_c, y_c) indicates the equation of clothoid to have the same radius of curvature as the circle.

$$X(t) = a \int_0^{\frac{a}{R\pi}} \cos\left(\frac{\pi}{2} u^2\right) du \tag{37}$$

$$Y(t) = a \int_0^{\frac{a}{R\pi}} \sin\left(\frac{\pi}{2} u^2\right) du \tag{38}$$

where t has been used in terms of a and R.

$$t = \frac{a}{R\pi} = \frac{ak}{\pi} \tag{39}$$

Equating the slope angle of the clothoid

$$\tan\theta = \tan\left(\frac{\pi t^2}{2}\right) = \tan\left(\frac{a^2}{2\pi R^2}\right) \tag{40}$$

and the slope angle of the circle

$$\tan\theta = -\frac{x - x_c}{y - y_c} \tag{41}$$

at a point y, we'll get

$$\tan\left(\frac{a^2}{2\pi R^2}\right) = -\frac{x - x_c}{y - y_c} \tag{42}$$

Searching for a match point in the right half of the circle

$$y - y_c = \sqrt{R^2 - (x - x_c)^2} \tag{43}$$

makes the slope equation to be a function of a

$$\tan\left(\frac{a^2}{2\pi R^2}\right)\sqrt{R^2 - \left(a\int_0^{\frac{a}{R\pi}} \cos\left(\frac{\pi}{2}u^2\right)\mathrm{d}u - x_c\right)^2} + a\int_0^{\frac{a}{R\pi}} \cos\left(\frac{\pi}{2}u^2\right)\mathrm{d}u - x_C = 0 \tag{44}$$

Solution of this equation provides us with an a for which the clothoid ends at a point with the same curvature as the circle. At the same y of the end point, the slope of the clothoid is also equal to the slope of the circle. A proper shift to the clothoid of the *X-axis* will match the clothoid and the circle. The following example will clarify the problem. Assume the given circle is

$$(x - 60)^2 + (y - 60)^2 = 50^2 \tag{45}$$

and therefore, the slope equation will be

$$\tan\left(\frac{a^2}{2\pi 2500}\right)\sqrt{2500 - \left(a\int_0^{\frac{a}{50\pi}} \cos\left(\frac{\pi}{2}u^2\right)\mathrm{d}u - 60\right)^2} + a\int_0^{\frac{a}{50\pi}} \cos\left(\frac{\pi}{2}u^2\right)\mathrm{d}u - 60 = 0 \tag{46}$$

The Numerical solution of the equation is

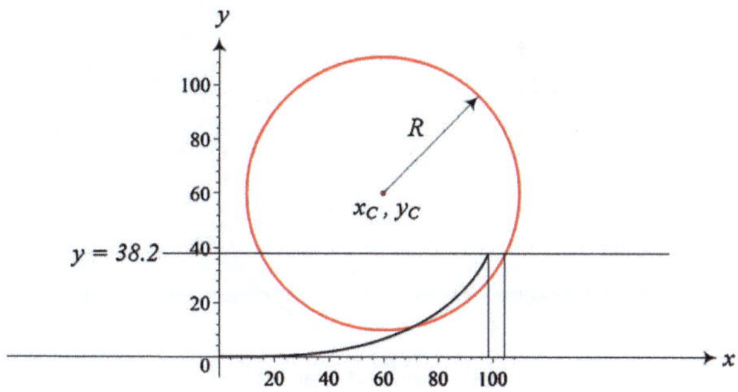

Fig. 7 A clothoid starting at the origin ends at a point with the same slope and curvature as the given circle, at the same *y*

$$a = 132.6477323 \tag{47}$$

The plot of the clothoid and the circle at this moment are shown in Fig. 7. For the calculated a, the value of t at the end point of the clothoid is

$$t = \frac{a}{R\pi} = \frac{2.652954646}{\pi} = 0.84446 \tag{48}$$

and therefore, the coordinates of the end point are

$$x(k) = 132.6 \int_0^{0.84} \cos\left(\frac{\pi}{2}u^2\right) \mathrm{d}u = 98.75389126 \tag{49}$$

$$y(k) = 132.6 \int_0^{0.84} \sin\left(\frac{\pi}{2}u^2\right) \mathrm{d}u = 38.22304651 \tag{50}$$

At the point, the radius of curvature of the clothoid is

$$R = \frac{1}{k} = \frac{a}{t\pi} = \frac{132.6477323}{0.84446\pi} = 50 \tag{51}$$

and the slope is

$$\theta = \frac{\pi}{2}t^2 = \frac{\pi}{2}0.84446^2 = 1.1202\,\mathrm{rad} \tag{52}$$

The x-coordinate of the circle at the same $y = 38.22304651$,

$$38.22304651 - 60 = \sqrt{50^2 - (x - 60)^2} \tag{53}$$

is

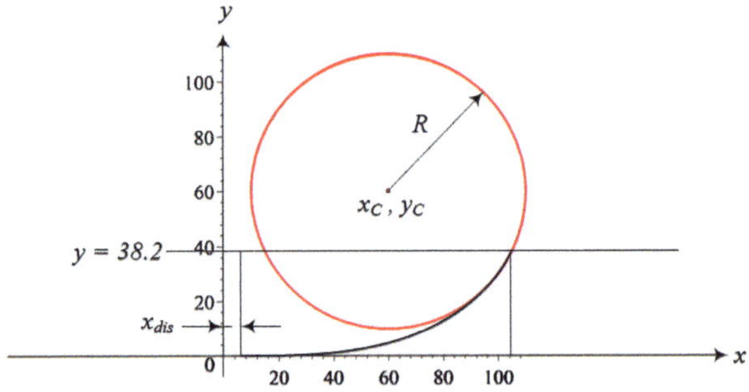

Fig. 8 A shifted clothoid starting on a point on the x-axis ends at a point with the same slope and curvature as the given circle and on the circle

$$x_{circle} = 105.0084914 \tag{54}$$

If we shift the clothoid by the difference between x_{circle} and $x_{clothoid}$

$$x_{dis} = x_{circle} - x_{clothoid} = 105.0084919 - 98.75389126 = 6.2546 \tag{55}$$

Then the clothoid and circle meet at a point on the circle with all requirements to have a smooth transition. Figure 8 illustrates the result.

3.2 Design Chart

Figure 9 illustrates a design graph of the relationship between the clothoid and parameters of magnification factor a, curvature k, and slope θ. The higher the magnification factor the larger the clothoid will be. The clothoid curves of different a are intersected by the constant slope lines of θ. The curves of constant curvature k intersect both, the constant a and constant θ curves. Assume we are trying to find a transition road with the clothoid shape to connect a straight road to a circle of radius $R = 58.824$ m. Having R is equivalent to have the destination curvature $k = 1/R = 0.017$. The desired circle must be tangent to a clothoid with a given a at the point that the clothoid is intersecting the curve of $k = 0.017$. The clothoid for $a = 250$ m hits the curve of $k = 0.017$ at a point for which we have

$$X = 147.3884878\,\text{m} \qquad Y = 176.4421850\,\text{m} \tag{56}$$

$$\theta = 164.7102491^\circ \qquad s = 338.2042540\,\text{m} \tag{57}$$

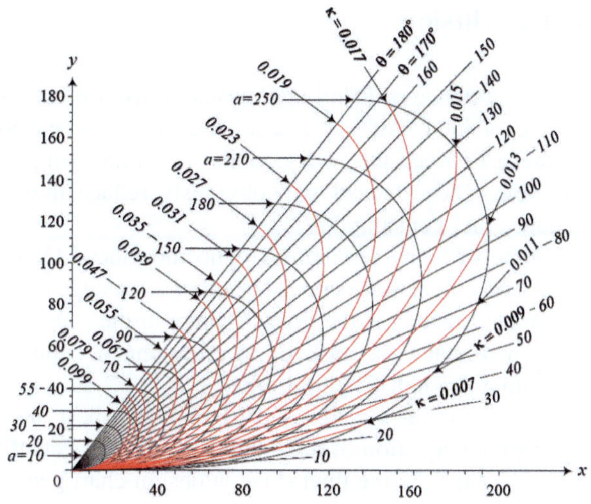

Fig. 9 A design graph of the relationship between the clothoid and parameters of magnification factor a, curvature k, and slope θ

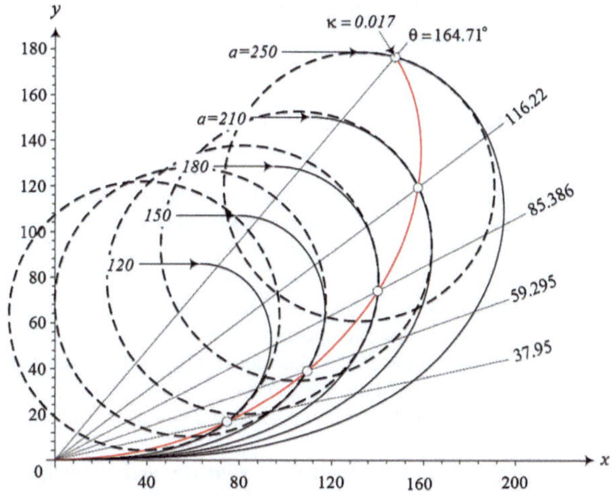

Fig. 10 A few clothoid transition road to connect a straight road to a circle of radius R = 58.824 m

The clothoid for $a = 210$ m hits the curve of $k = 0.017$ at a point for which we have

$$X = 157.4739501\,\text{m} \qquad Y = 119.7133227\,\text{m} \tag{58}$$

$$\theta = 116.2195518^\circ \qquad s = 238.6369216\,\text{m} \tag{59}$$

and so on. Figure 10 illustrates these solutions and some more. The number of solutions is practically infinite and the best solution depends on safety cost, and physical constraints of the field [5–9].

4 Conclusion

The importance of using appropriate transition curves, for sustainable road design was explained. An appropriate transition curve should help with reducing the distance between the two sections which are being connected in the shortest time possible. These factors will obviously reduce the need for higher speeds which reduces fuel consumption.

Some of many researchers, who introduced Clothoids as the best option for a transition curves were cited. The other advantage of the clothoids besides the above mentioned factors is the linear change in the steering angle required by the driver. This will make the task of steering much simpler and more consistent while travelling through Clothoid turns.

Some examples were used to show the best and easiest ways for finding the appropriate equation of a Clothoid which connects different types of road sections to each other. As the final stage, a design chart generated as a result of calculation for many different scenarios was introduced. This design chart can be used to find appropriate numerical values which are required to be substituted into the equation of a Clothoid with respect to the sections on its both sides.

References

1. Komoriya K, Tanie K (1989) Trajectory design and control of a wheel-type mobile robot using B-spline curve. In: IEEE/RSJ international workshop on intelligent robots and systems '89. The autonomous mobile robots and its applications. IROS '89 Proceedings, pp 398–405, 4–6 Sep 1989
2. Kanayama Y, Miyake N (1985) Trajectory generation of mobile robots. In: Robotics research: the third international symposium. ISBN: 0262061015
3. Lamm R, Psarianos B, Mailaender T (1999) Highway design and traffic safety engineering handbook. McGraw-Hill, New York
4. Talbot AN (1899) The railway transition spiral. U. Illinois. Reprinted from the Technograph no 13
5. Jazar RN (2014) Vehicle dynamics: theory and application, 2nd edn. Springer, New York
6. Marzbani H, Jazar R, Fard M (2013) Hydraulic engine mounts: a survey. J Vib Control 19(16):2
7. Marzbani H, Jazar RN, Khazaei A (2012) Smart passive vibration isolation: requirements and unsolved problems. J Appl Nonlinear Dyn 1(4):341–386
8. Marzbani H, Jazar RN (2014) Smart flat ride tuning. Nonlinear approaches engineering applications, vol 2. Springer, New York, pp 3–36
9. Marzbani H (2012) Flat ride; problems and solutions in vehicle. Nonlinear Eng 1(3–4):101–108

Sustainable Flat Ride Suspension Design

Hormoz Marzbani, Reza N. Jazar and M. Fard

Abstract Maurice Olley suggested that having natural frequency of the front approximately 80 % of that of the rear suspension in a vehicle will result in a flat ride for the passengers. Flat Ride in this case means that the pitch motion of the vehicle, generated by riding over a bump for instance will fade into the bounce motion of the vehicle much faster. Bounce motion of the vehicle in mush easier to tolerate and feels more comfortable for the passengers. In a previous study which is shortly mentioned in this paper the authors, analytically proved that this situation is not practical. In other words, for any vehicle there will only be one certain velocity, depending on the geometry and suspension system specifications which the flat ride will happen at. The search continued to find a practical method for enjoying the flat ride in vehicles. Solving the equation of motion of the vehicle for different spring rates and situation the authors came up with design chart for smart suspension systems. Using the advantages of the analytical approach to the flat ride problem, the chart was established to be used for vehicles with smart active suspension systems. In this paper the mathematical methods used and the resulted criteria for designing a flat ride suspension system which will perform in different speeds is presented.

Keywords Flat ride · Maurice Olley · Optimal suspension · Vehicle vibrations · Vehicle dynamics · Suspension design · Suspension optimization

H. Marzbani (✉) · R.N. Jazar · M. Fard
School of Aerospace, Mechanical and Manufacturing Engineering, RMIT University, Melbourne, Australia
e-mail: hormoz.marzbani@rmit.edu.au

R.N. Jazar
e-mail: reza.jazar@rmit.edu.au

M. Fard
e-mail: mohammad.fard@rmit.edu.au

I. Denbratt et al. (eds.), *Sustainable Automotive Technologies 2014*,
Lecture Notes in Mobility, DOI 10.1007/978-3-319-17999-5_4

1 Introduction

The excitation inputs from the road to a straight moving car will affect the front wheels first and then, with a time lag, the rear wheels. The general recommendation was that the natural frequency of the front suspension should be lower than that of the rear. So, the rear part oscillates faster to catch up with the front to eliminate pitch and put the car in bounce before the vibrations die out by damping. This is what Olley called the *Flat Ride Tuning* [1]. Maurice Olley (1889–1983) established guidelines, back in the 1930, for designing vehicles with better ride. These were derived from experiments with a modified car to allow variation of the pitch mass moment. Although the measures of ride were strictly subjective, those guidelines are considered as valid rules of thumb even for modern cars. What is known as Olley's Flat Ride not considering the other prerequisites can be put forward as:

> The front suspension should have around 30 % lower rate than the rear.

An important prerequisite for flat ride was the uncoupling condition, which was introduced by Rowell and Guest for the first time in 1923 [1]. Rowell and Guest used the geometry of a bicycle car model to find the condition which sets the bounce and pitch centers of the model located on the springs. Having the condition, the front and rear spring systems of the vehicle can be regarded as two separate one degree of freedom systems.

What happens as a result of applying the Olley's conditions to a car is the fact that the pitch motion of the vehicle will turn into a bounce motion. Pitch motion provides a much more uncomfortable experience for the passengers compared to bounce motion.

As mentioned earlier ride comfort is a personal experience which is different from one individual to other, but the effects of motion can be similar. Motion in the car can cause fatigue for the passengers. Driver fatigue is a significant cause of accidents on motorways. The fatigue caused by driving extended periods actually impairs driver alertness and performance and therefore can compromise transportation safety [2].

Whole body vibration has been found to correlate with a range of psychological reactions of the human body such as lower back pain and heart rate variability [3, 4]. Disturbance of vision and balance have also been reported to occur [5].

So a better suspension design, will have an enormous effect on the ride comfort, which will result in a safer transportation as well as a more comfortable travelling experience.

In this study, using analytical methods, we study the flat ride conditions which has been respected and followed by the car manufacturers' designers since they were introduced for the first time. This article will provide a more reliable scientific and mathematical approach for what are the flat ride design criteria in vehicle dynamic studies.

2 Previous Works

Maurice Olley was one of the first pioneers who introduced and studied the concept of flat ride in vehicle dynamics. He was an English engineer born in 1889, who during his life added a lot to the general knowledge of vehicle dynamics and is counted as one of the great automobile engineers of his era. He is one of the founders of modern vehicle dynamics. In his early career in the Rolls Royce design office, he worked under Sir Henry Royce but the majority of his career was spent at Cadillac in the USA and Vauxhall in England.

Olley worked directly for Sir Henry Royce, and was in the United States for some 10 years struggling to get off the ground the manufacture of Rolls-Royce cars at Springfield, Massachusetts. The financial crash of 1929 put the skids under the operation. His first task after moving to the Cadillac company in 1930 was suspension and ride. He introduced the Rolls-Royce type of bump rig and began a full program of ride development. He studied the oscillation of wheels and tires and by applying some changes on the rig was soon studying the basic ride motions of the car. In his paper [6] he published the results taken from his experiments using the test rig for the first time.

He developed a bouncing table rig in General Motors proving grounds, on which humans were vibrated vertically at different frequencies and amplitudes. They would have increased the frequency till the person on the table begins to feel uncomfortable. Using this equipment Olley explained the relation between vertical acceleration and comfort over a range of frequencies. He generated a curve for passenger comfort, which is very similar to the current *ISO*2631 standard.

Olley as well as other investigators in well-established car companies realized that the pitch and roll modes of the car body are much more uncomfortable than the bounce mode. The investigators' effort focused on the suspension stiffness and damping rates to be experimentally adjusted to provide acceptable vertical vibrations. However, the strategy about roll and pitch modes were to transform them to bounce. Due to usual geometric symmetry of cars, as well as the symmetric excitation from the road, roll mode are being excited much less than pitch mode. Therefore, lots of investigations have been focused on adjustment of the front and rear suspensions such that pitch mode of vibration transform to the bounce.

In the early 1930 s most cars were built with fairly stiff springs at the front and soft at the rear, with a $\frac{r^2}{a_1 a_2}$ ratio in pitch of about 0.8, where r is the pitch radius of gyration of the car and a_1 and a_2 are the distance of the mass center, C, from the front and rear axles, as shown in Fig. 1, [7]. However, based on what Olley discovered, such a choice was against the mode transfer desire.

Besides all the important facts that Olley discovered during his experiments, the principle known as the Flat Ride Tuning or Olley's Flat Ride proved to be more industry approved and accepted. After his publications [6, 8, 9] in which he advocated this design practice, they became rules of practice.

We can summarize what has been said about ride and comfort in American passenger cars by Olley as the following:

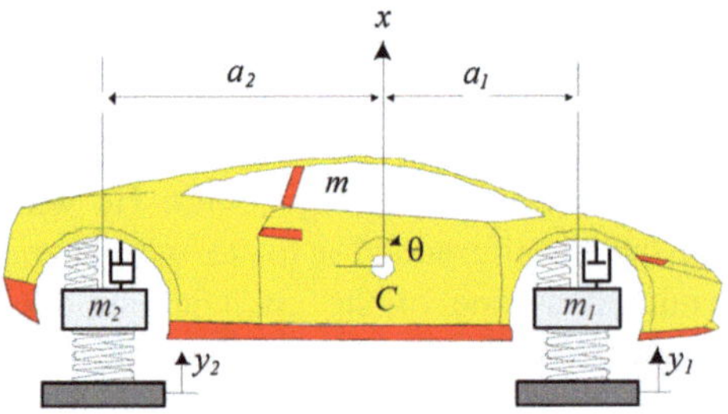

Fig. 1 Bicycle car model used for vibration analysis

1. The front spring should be softer than the rear for Flat Ride Tuning. This will promote bouncing of the body rather than pitching motions at least for a greater majority of speeds and bump road situations. The front suspension should have a 30 % lower ride rate than the rear suspension, or the spring center should be at least 6.5 % of the wheelbase behind the center of gravity. Although this does not explicitly determine the front and rear natural frequencies, since the front-rear weight distribution on passenger cars is close to 50–50, it will generally assure that the rear frequency is greater than the front.
2. The ratio $\frac{r^2}{a_1 \cdot a_2}$ normally approaches unity. This reduces vibration interactions between front and rear because the two suspensions can now be considered as two separate systems. As a consequence there will be less resonant build-ups on the road and the pitching frequency will have a magnitude closer to that of bounce.
3. The pitch and bounce frequencies should be close together: the bounce frequency should be less than 1.2 times the pitch frequency. For higher ratios interference kicks resulting from the superposition of the two motions. This condition will be met for modern cars because the dynamic index is near unity with the wheels located near the forward and rearward extremes of the chassis.
4. Neither frequency should be greater than 1.3 Hz, which means that the effective static deflection of the vehicle should exceed roughly 6 in.
5. The roll frequency should be approximately equal to the pitch and bounce frequencies. To minimize roll vibrations the natural frequency in roll needs to be low just as for the bounce and pitch modes.

Rowell and Guest [10] in 1923 identified the value of $\frac{r^2}{a_1 \cdot a_2}$ being associated with vehicles in which the front and rear responses were uncoupled. Olley was able to investigate the issue experimentally and these experiments led him to the belief that pitching motion was extremely important in the subjective assessment of vehicle ride comfort. He built the Cadillac k_2 rig in 1931 which was a 12 cylinder, 7 passenger Cadillac limousine of the period, fitted with front and rear outriggers each of which could carry up to 327 kg made up in 27 kg weights. To their surprise, under these supposedly ideal conditions, they still got an unsatisfactory ride. This arrangement gave no fixed oscillation centers and the ride had no pattern. However, by fitting all the weights they found that if the front spring static deflections are some 30 % greater than the rear then the revolutionary flat ride occurs. Olley's explanation was that because the two ends of the car did not cross a given disturbance at the same instant it was important that the front wheels initiated the slower mode and that the rear wheels initiated the faster mode. This allowed the body movement at the rear to catch up the front and so produce the flat ride.

The condition of Flat Ride is expressed in various detailed forms; however, the main idea states that *the front suspension should have a* 30 % *lower ride rate then the rear.* The physical explanation for why this is beneficial in reducing pitch motion is usually argued based on the time history of events following a vehicle hitting a bump. First, the front of the vehicle responds approximately in the well-known damped oscillation manner. At some time later, controlled by the wheelbase and the vehicle speed, the rear responds in similar fashion. The net motion of the vehicle is then crudely some summation of these two motions which minimizes the vehicle pitch response, [11].

Confirmation of the effectiveness in pitch reduction of the Olley design was given by Best [7] over a limited range of circumstances. Random road excitation was applied to a half-car computer model, with identical front and rear excitations, considering the time delay generated by the wheelbase and vehicle's speed. Pitch suppression was associated with the wheelbase filtering effect. Pitch suppression appeared to be necessarily associated with increases in bounce response, leaving in unclear whether or not it is a worthwhile goal, [12].

Sharp and Pilbeam [13] attempted a more fundamental investigation of the phenomenon, primarily by calculating frequency response for the half-car over a wide range of speed and design conditions. At higher speeds, remarkable reductions in pitch response with only small costs in terms of bounce response were shown. At low speeds, the situation is reversed. These behavioral features were shown to be generic insofar as variations in mass center location, pitch inertia and damping level were concerned, and the implifications from the frequency responses were confirmed by simulations with nonlinear asymmetric suspension damping.

Later on Sharp [12] discussed the rear to front stiffness tuning of the suspension system of a car, through reference to a half-car pitch plane mathematical model. He used new results relating to the frequency responses of the bouncing and pitching motions of the car body to show that the pitch minimization mechanism of Olley's Flat Ride tuning involves interference between the responses to the front and rear axle inputs. He showed that interference with respect to the rotational motion implies reinforcement with respect to the translational motion, and vice versa. Sharp conclude almost the same facts mentioned by Best and other researchers before him, saying that at higher vehicle speeds, Olley tuning is shown to bring advantage in pitch suppression with a very little disadvantage in terms of body acceleration. At lower speeds, he continues, not only does the pitch tuning bring large vertical acceleration penalties but also suspension stiffness implied are impractical from an attitude control standpoint.

The flat ride problem was revisited by Crolla and King [11]. They generated vehicle vibration response spectra under random road excitations. Some results included the wheelbase filter effect, while others did not. Olley and reverse Olley designs were simulated at speeds of 10, 20, 30, and 40 m/s, with the result that

Olley design was good in pitch and bad in bounce in all cases. It was confidently concluded that the rear/front stiffness ratio has virtually no effect on overall levels of ride comfort.

In 2004, Odhams and Cebon investigated the tuning of a pitch-plane model of a passenger car with a coupled suspension system and compared it to that of a conventional suspension system, which followed the Rowell and Guest treatment, [14]. They believed that there is a significant benefit from coupling front and rear suspensions; coupled suspensions with a "Hydrolastic" or "Hydragas" systems, in which the front and rear suspension struts are connected hydraulically, have proved very effective in some applications. The concluded that the Olley's flat ride tuning provides a near optimum stiffness choice for conventional suspensions for minimizing dynamic tire forces and is very close to optimal for minimizing horizontal acceleration at the chest (caused by pitching) but not the vertical acceleration.

3 Uncoupling the Car Bicycle Model

Consider the two degree-of-freedom (*DOF*) system in Fig. 2. A beam with mass m and mass moment I about the mass center C is sitting on two springs k_1 and k_2 to model a car in bounce and pitch motions. The translational coordinate x of C and the rotational coordinate θ are the usual generalized coordinates that we use to measure the kinematics of the beam. The equations of motion and the mode shapes are functions of the chosen coordinates.

The free vibration equations of motion of the system are:

$$\begin{bmatrix} m & 0 \\ 0 & I \end{bmatrix} \begin{bmatrix} \ddot{x} \\ \ddot{\theta} \end{bmatrix} + \begin{bmatrix} k_1 + k_2 & a_2 k_2 - a_1 k_1 \\ a_2 k_2 - a_1 k_1 & a_2^2 k_2 + a_1^2 k_1 \end{bmatrix} \begin{bmatrix} x \\ \theta \end{bmatrix} = 0 \tag{1}$$

To compare the mode shapes of the system practically, we employ the coordinates x_1 and x_2 instead of x and θ. The equations of motion of the system would then be:

$$\begin{bmatrix} \frac{ma_2^2+I}{a_1+a_2^2} & \frac{ma_1a_2-I}{a_1+a_2^2} \\ \frac{ma_1a_2-I}{a_1+a_2^2} & \frac{ma_1^2+I}{a_1+a_2^2} \end{bmatrix} \begin{bmatrix} \ddot{x}_1 \\ \ddot{x}_2 \end{bmatrix} + \begin{bmatrix} k_1 & 0 \\ 0 & k_2 \end{bmatrix} \begin{bmatrix} x_1 \\ x_2 \end{bmatrix} = 0 \tag{2}$$

Let us define the following parameters:

$$I = mr^2 \tag{3}$$

$$\Omega_1^2 = \frac{k_1}{m}\beta \tag{4}$$

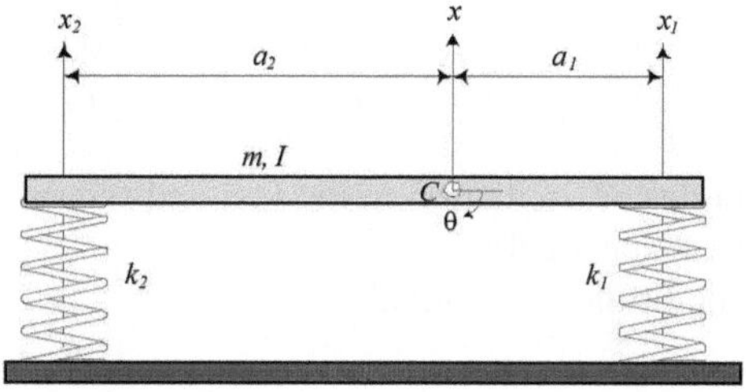

Fig. 2 The bicycle model of a car is a beam of mass m and mass moment I, sitting on two springs k_1 and k_2

$$\Omega_2^2 = \frac{k_2}{m}\beta \tag{5}$$

$$\beta = \frac{l^2}{a_1 a_2} \tag{6}$$

$$\alpha = \frac{r^2}{a_1 a_2} \tag{7}$$

$$\gamma = \frac{a_2}{a_1} \tag{8}$$

$$l = a_1 + a_2 \tag{9}$$

and rewrite the equations as

$$\begin{bmatrix} \alpha+\gamma & 1-\alpha \\ 1-\alpha & \alpha+\frac{1}{\gamma} \end{bmatrix} \begin{bmatrix} \ddot{x}_1 \\ \ddot{x}_2 \end{bmatrix} + \begin{bmatrix} \Omega_1^2 & 0 \\ 0 & \Omega_2^2 \end{bmatrix} \begin{bmatrix} x_1 \\ x_2 \end{bmatrix} = 0 \tag{10}$$

Setting

$$\alpha = 1 \tag{11}$$

makes the equations decoupled

$$\begin{bmatrix} \alpha+\gamma & 0 \\ 0 & \alpha+\frac{1}{\gamma} \end{bmatrix} \begin{bmatrix} \ddot{x}_1 \\ \ddot{x}_2 \end{bmatrix} + \begin{bmatrix} \Omega_1^2 & 0 \\ 0 & \Omega_2^2 \end{bmatrix} \begin{bmatrix} x_1 \\ x_2 \end{bmatrix} = 0 \tag{12}$$

The natural frequencies ω_i and mode shapes u_i of the system are

$$\omega_1^2 = \frac{1}{\gamma+1}\Omega_1^2 = \frac{l}{a_2}\frac{k_1}{m} \qquad u_1 = \begin{bmatrix} 1 \\ 0 \end{bmatrix} \tag{13}$$

$$\omega_2^2 = \frac{\gamma}{\gamma+1}\Omega_2^2 = \frac{l}{a_1}\frac{k_2}{m} \quad u_2 = \begin{bmatrix} 0 \\ 1 \end{bmatrix} \tag{14}$$

They show that the nodes of oscillation in the first and second modes are at the rear and front suspensions respectively.

The decoupling condition $\alpha = 1$ yields

$$r^2 = a_1 a_2 \tag{15}$$

which indicates that the pitch radius of gyration, r, must be equal to the multiplication of the distance of the mass canter C from the front and rear axles. Therefore, by setting $\alpha = 1$, the nodes of the two modes of vibrations appear to be at the front and rear axles. As a result, the front wheel excitation will not alter the body at the rear axle and vice versa. For such a car, the front and rear parts of the car act independently. Therefore, the decoupling condition $\alpha = 1$ allows us to break the initial two *DOF* system into two independent one *DOF* systems, where:

$$m_r = m\frac{a_1}{l} = m\varepsilon \tag{16}$$

$$m_f = m\frac{a_2}{l} = m(1-\varepsilon) \tag{17}$$

$$\varepsilon = \frac{a_1}{l} \tag{18}$$

The equations of motion of the independent systems will be:

$$m(1-\varepsilon)\ddot{x_1} + c_1\dot{x_1} + k_1 x_1 = k_1 y_1 + c_1 \dot{y}_1 \tag{19}$$

$$m\varepsilon\ddot{x_2} + c_2\dot{x_2} + k_2 x_2 = k_2 y_2 + c_2\ddot{y_2} \tag{20}$$

The decoupling condition of undamped free system will not necessarily decouple the general damped system. However, if there is no anti-pitch spring or anti-pitch damping between the front and rear suspensions then equations of motion

$$\begin{bmatrix} \alpha+\gamma & 1-\alpha \\ 1-\alpha & \alpha+\frac{1}{\gamma} \end{bmatrix}\begin{bmatrix} \ddot{x_1} \\ \ddot{x_2} \end{bmatrix} + \begin{bmatrix} 2\xi_1\Omega_1 & 0 \\ 0 & 2\xi_2\Omega_2 \end{bmatrix}\begin{bmatrix} \dot{x_1} \\ \dot{x_2} \end{bmatrix} + \begin{bmatrix} \Omega_1^2 & 0 \\ 0 & \Omega_2^2 \end{bmatrix}\begin{bmatrix} x_1 \\ x_2 \end{bmatrix}$$
$$= \begin{bmatrix} 2\xi_1\Omega_1 & 0 \\ 0 & 2\xi_2\Omega_2 \end{bmatrix}\begin{bmatrix} \dot{y_1} \\ \dot{y_2} \end{bmatrix} + \begin{bmatrix} \Omega_1^2 & 0 \\ 0 & \Omega_2^2 \end{bmatrix}\begin{bmatrix} y_1 \\ y_2 \end{bmatrix} \tag{21}$$

$$2\xi_1\Omega_1 = \frac{c_1}{m}\beta \tag{22}$$

$$2\xi_2\Omega_2 = \frac{c_2}{m}\beta \tag{23}$$

will be decoupled by $\alpha = 1$.

$$\begin{bmatrix} \alpha+\gamma & 0 \\ 0 & \alpha+\frac{1}{\gamma} \end{bmatrix}\begin{bmatrix} \ddot{x}_1 \\ \ddot{x}_2 \end{bmatrix} + \begin{bmatrix} c_1 & 0 \\ 0 & c_2 \end{bmatrix}\begin{bmatrix} \dot{x}_1 \\ \dot{x}_2 \end{bmatrix} + \begin{bmatrix} \Omega_1^2 & 0 \\ 0 & \Omega_2^2 \end{bmatrix}\begin{bmatrix} x_1 \\ x_2 \end{bmatrix} = \begin{bmatrix} 2\xi_1\Omega_1 & 0 \\ 0 & 2\xi_2\Omega_2 \end{bmatrix}\begin{bmatrix} \dot{y}_1 \\ \dot{y}_2 \end{bmatrix} + \begin{bmatrix} \Omega_1^2 & 0 \\ 0 & \Omega_2^2 \end{bmatrix}\begin{bmatrix} y_1 \\ y_2 \end{bmatrix} \tag{24}$$

The equations of motion of the independent system may also be written as

$$m(1-\varepsilon)\ddot{x}_1 + c_1\dot{x}_1 + k_1x_1 = c_1\dot{y}_1 + k_1y_1 \tag{25}$$

$$m\varepsilon\ddot{x}_2 + c_2\dot{x}_2 + k_2x_2 = c_2\dot{y}_2 + k_2y_2 \tag{26}$$

which are consistent with the decoupled Eq. 36 because of

$$\varepsilon = \frac{1+\gamma}{\gamma\Omega_2^2} \tag{27}$$

4 No Flat Ride Solution for Linear Suspension

The time lag between the front and rear suspension oscillations is a function of the wheelbase, l, and speed of the vehicle, υ. Soon after the rear wheels have passed over a step, the vehicle is at the worst condition of pitching. Olley experimentally determined a recommendation for the optimum frequency ratio of the front and rear ends of cars. His suggestion for American cars and roads of 50 s was to have the natural frequency of the front approximately 80 % of that of the rear suspension.

To examine Olley's experimental recommendation and possibly make an analytical base for flat ride, let us rewrite the equation of motion (25) and (26) as:

$$\ddot{x}_1 + 2\xi_1\dot{x}_1 + \frac{k_1}{m(1-\varepsilon)}x_1 = 2\xi_1\dot{y}_1 + \frac{k_1}{m(1-\varepsilon)}y_1 \tag{28}$$

$$\ddot{x}_2 + 2\xi\xi_1\dot{x}_2 + \frac{kk_1}{m\varepsilon}x_2 = 2\xi\xi_1y_2 + \frac{kk_1}{m\varepsilon}y_2 \tag{29}$$

where,

$$\xi = \frac{\xi_2}{\xi_1} = \frac{c_1}{c_2} \frac{\varepsilon}{1-\varepsilon} \tag{30}$$

$$k = \frac{k_2}{k_1} = \frac{k_1}{k_2} \frac{\varepsilon}{1-\varepsilon} \tag{31}$$

$$\xi_1 = \frac{c_1}{m(1-\varepsilon)} \tag{32}$$

$$\xi_2 = \frac{c_2}{m\varepsilon} \tag{33}$$

Parameters k and ξ are the ratio of the rear/front spring rates and damping ratios respectively.

The necessity to achieve a flat ride provides that the rear system must oscillate faster to catch up with the front system at a reasonable time. At the time both systems must be at the same amplitude and oscillate together afterwards. Therefore, an ideal flat ride happens if the frequency of the rear system be higher than the front to catch up with the oscillation of the front at a certain time and amplitude. Then, the frequency of the rear must reduce to the value of the front frequency to oscillate in phase with the front. Furthermore, the damping ratio of the rear must also change to keep the same amplitude. Such a dual behavior is not achievable with any linear suspension. Therefore, theoretically, it is impossible to design linear suspensions to provide a flat ride, as the linearity of the front and rear suspensions keep their frequency of oscillation constant.

5 Nonlinear Damper

The force-velocity characteristics of an actual shock absorber can be quite complex. Although we may express the complex behavior using an approximate function, analytic calculation can be quite complicated with little design information. Furthermore, the representations of the exact shock absorber do not greatly affect the behavior of the system. The simplest linear viscous damper model is usually used for linear analytical calculation

$$F_D = c v_D \tag{34}$$

where c is the damping coefficient of the damper.

The bound and rebound forces of the damper are different, in other words the force-velocity characteristics diagram is not symmetric. Practically, a shock

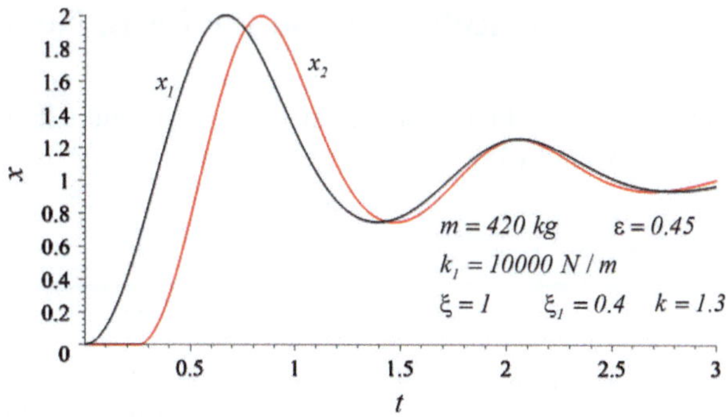

Fig. 3 Response of the front and rear suspension of a near flat ride car with ideal nonlinear damper to a unit step input

absorber compresses much easier than decompression. A reason is that during rebound in which the damper extends back, it uses up the stored energy in the spring. A high compression damping, prevents to have enough spring compression to collect enough potential energy. That is why in order to get a more reliable and close to reality response for analysis on dampers, using bilinear dampers is suggested. It is similar to a linear damper but with different coefficients for the two directions (Dixon 2008).

$$F_D = \begin{cases} c_{DE}\upsilon_D & \textit{Extension} \\ c_{DC}\upsilon_D & \textit{Compression} \end{cases} \tag{35}$$

where c_{DE} is the damping coefficient when damper is extended and c_{DC} is the damping coefficient when the damper is compressed.

An ideal dual behavior damper is one which does not provide any damping while being compressed and on the other hand damps the motion while extending.

After using the nonlinear model for the damper, the motion had to be investigated in 3 steps for the front and same for the rear. Ideally, the unit step moves the ground up in no time and therefore the motion of the system begins when $y = 1$ and the suspension is compressed. The first step is right after the wheel hits the step and the damper starts extending. The second step is when the damper starts the compression phase, which the damping coefficient would be equal to zero. The third step is when the damper starts extending again. Each of the equations of motion should be solved for the 3 steps separately in order to find the time and amplitude of the third peak of the motion which have been chosen to be optimal time for the flat ride to happen at.

Figure 3 illustrates the behavior of the car equipped with a nonlinear damper when going over a unit step input.

6 Near Flat Ride Solution for Ideal, Nonlinear Damper

The conditions that x_1 and x_2 meet after one and a half oscillations can be shown by Eqs. (36) and (37).

$$x_1 = x_2 \tag{36}$$

$$t_{p1} = t_{p2} \tag{37}$$

The equation resulted from $x_1 = x_2$ (Eq. 38), has got ξ and ξ_1 as its variables and could be plotted as an explicit function of the variables which interestingly shows that the value for $\xi = \xi_2/\xi_1$ must equal to 1 for any value for damping coefficient of the front suspension ξ_1.

$$\begin{aligned} EQ1 = &\frac{-0.8 \times 10^{-18}}{(\zeta_1^2 - 1)(\zeta_1^2\zeta^2 - 1)}(-0.3172834025 \times 10^{18} e^{\frac{-\pi\zeta_1}{\sqrt{1-\zeta_1^2}}}\zeta_1^4\zeta^2 \\ &+ 0.3172834025 \times 10^{18} e^{\frac{-\pi\zeta_1}{\sqrt{1-\zeta_1^2}}}\zeta_1^2 + 0.3172834025 \times 10^{18} e^{\frac{-\pi\zeta_1}{\sqrt{1-\zeta_1^2}}}\zeta_1^2\zeta^2 \\ &- 0.3172834025 \times 10^{18} e^{\frac{-\pi\zeta_1}{\sqrt{1-\zeta_1^2}}} \\ &+ 0.130151797 \times 10^{9} e^{\frac{-\pi\zeta_1}{\sqrt{1-\zeta_1^2}}}\sqrt{1 - \zeta_1^2\zeta_1^3\zeta^2} \\ &- 0.3172834025 \times 10^{9} e^{\frac{-\pi\zeta_1}{\sqrt{1-\zeta_1^2}}}\sqrt{1 - \zeta_1^2\zeta_1} \\ &+ 0.3172834025 \times 10^{18} e^{\frac{-\pi\zeta_1\zeta}{\sqrt{1-\zeta_1^2\zeta^2}}} - 0.3172834025 \times 10^{18} e^{\frac{-\pi\zeta_1\zeta}{\sqrt{1-\zeta_1^2\zeta^2}}}\zeta_1^2\zeta^2 \\ &- 0.3172834025 \times 10^{18} e^{\frac{-\pi\zeta_1\zeta}{\sqrt{1-\zeta_1^2\zeta^2}}}\zeta_1^2 + 0.3172834025 \times 10^{18} e^{\frac{-\pi\zeta_1\zeta}{\sqrt{1-\zeta_1^2\zeta^2}}} \\ &- 0.130151797 \times 10^{9} e^{\frac{-\pi\zeta_1\zeta}{\sqrt{1-\zeta_1^2\zeta^2}}}\sqrt{1 - \zeta_1^2\zeta^2\zeta_1^3\zeta} \\ &+ 0.130151797 \times 10^{9} e^{\frac{-\pi\zeta_1\zeta}{\sqrt{1-\zeta_1^2\zeta^2}}}\sqrt{1 - \zeta_1^2\zeta^2\zeta_1\zeta}) \end{aligned} \tag{38}$$

Therefore, regardless of the value of ξ_1 the rear suspension should have an equal coefficient for the damper. The equation resulted from $t_{P_1} = t_{p_2}$ generates Eq. 39 to determine $k = k_2/k_1$. Figure 4 illustrates the spring ratio $k = k_2/k_1$ versus $\tau = l/\upsilon$, to have near flat ride with ideal nonlinear damping, for different $\varepsilon = a_1/l$.

$$EQ2 = \frac{\pi\sqrt{\frac{-k_1}{m(\varepsilon-1)}(2\sqrt{1-\zeta_1^2}+1-\zeta_1^2)m(\varepsilon-1)}}{(\zeta_1^2-1)k_1} - \tau + \frac{\pi(2\sqrt{1-\zeta^2\zeta_1^2}+1-\zeta^2\zeta_1^2)}{\sqrt{\frac{kk_1}{m\varepsilon}}(\zeta^2\zeta_1^2-1)} \tag{39}$$

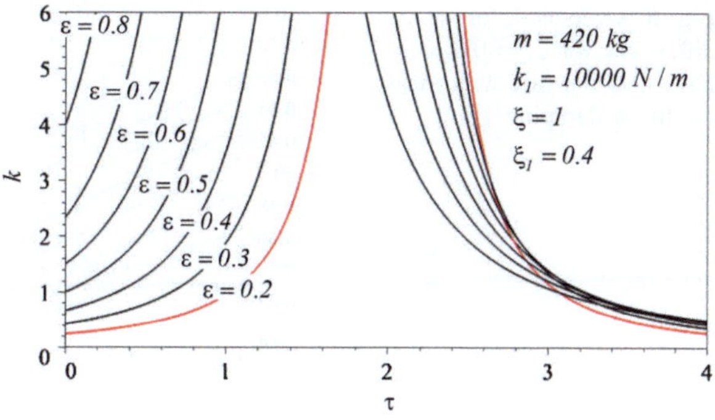

Fig. 4 The value of spring ratio $k = k_2/k_1$ versus $\tau = l/v$ to have near flat ride with ideal nonlinear damping, for different $\varepsilon = a_1/l$

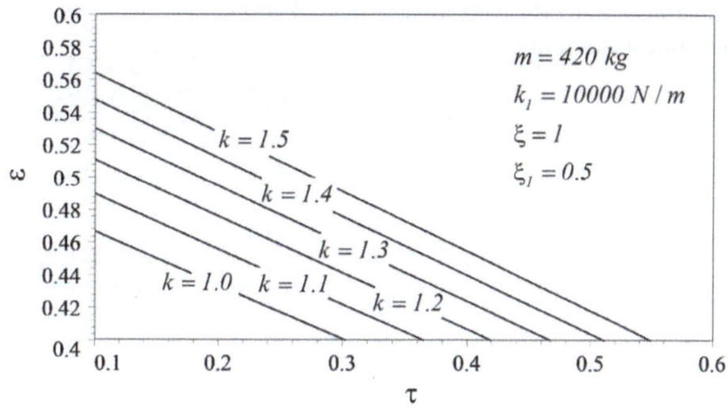

Fig. 5 ε versus τ, for different k for $\xi_1 = 0.5$ to have near flat ride with ideal nonlinear damping

The average length of a sedan vehicle has been taken equal to 2.6 m with a normal weight distribution of a front differential vehicle 56/44 heavier at the front. Using the given information some other values can be calculated as: $a_1 = 1144$ mm and $a_2 = 1456$ mm which yields to $\varepsilon = 0.44$. Considering the existing designs for street vehicles, only the small section of $0.1 < \tau < 0.875$ is applied. The mass center of street cars is also limited to $0.4 < \varepsilon < 0.6$.

Figure 5 shows how k varies with τ for $\xi_1 = 0.5$ and different ε to provide a near flat ride with ideal nonlinear damper. For any ε, the required stiffness ratio increases by increasing τ. Therefore, the ratio of rear to front stuffiness increases when the speed of the car decreases. Figures 6, 7, 8 and 9 also provide the same design graphs for $\xi_1 = 0.4$, 0.3, 0.2, 0.1, respectively.

There will be a possibility of using the τ versus ε diagrams as a design chart, which has been illustrated using the values in Table 1, by Fig. 10.

The box in Fig. 10, is indicating the values that the spring rate should be having as the travelling speed of the vehicle changes to provide the passengers with a flat ride, in case of having a smart active suspension. The point on the figure is an example for a passive suspension vehicle. It is showing the required spring rate, for getting a flat ride in a car with a wheelbase of 2.6 m, traveling at 28 km/h.

Fig. 6 ε versus τ, for different k for $\xi_1 = 0.4$ to have near flat ride with ideal nonlinear damping

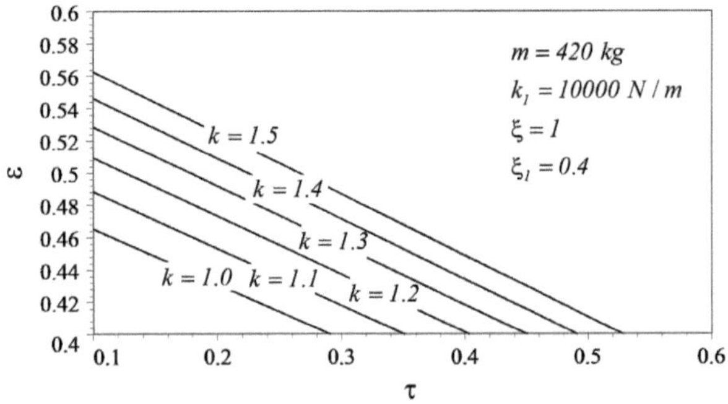

Fig. 7 ε versus τ, for different k for $\xi_1 = 0.3$ to have near flat ride with ideal nonlinear damping

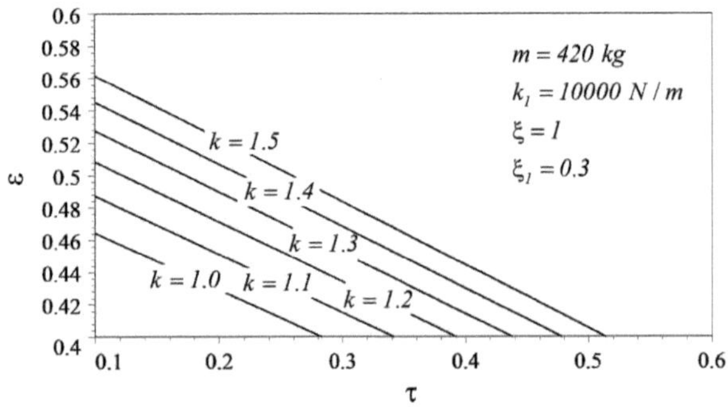

Fig. 8 ε versus τ, for different k for $\xi_1 = 0.2$ to have near flat ride with ideal nonlinear damping

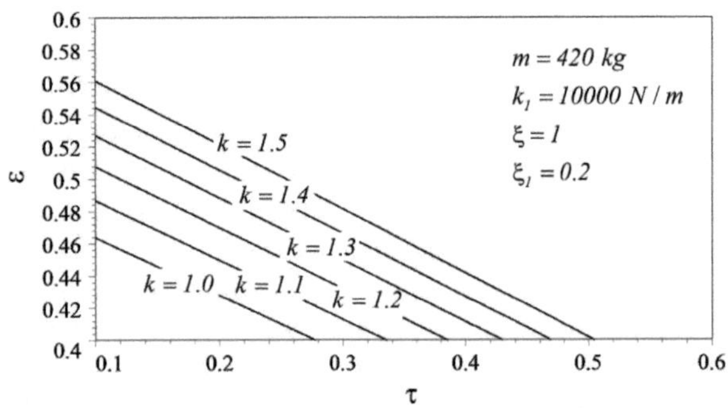

Fig. 9 ε versus τ, for different k for $\xi_1 = 0.1$ to have near flat ride with ideal nonlinear damping

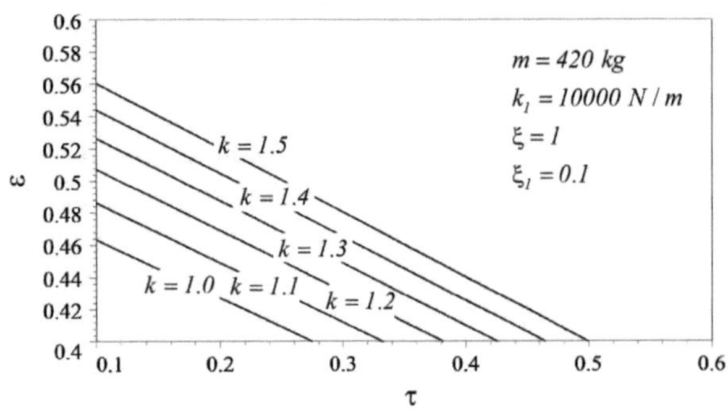

Table 1 Specification of a sample car

Specification	Nominal value
m (kg)	420
a_1 (m)	1.4
a_2 (m)	1.47
l (m)	2.87
k_1 (N/m)	10,000
k_2 (N/m)	13,000
c_1 (N s/m)	1000
c_2 (N s/m)	1000
β	4.00238
γ	1.05
Ω_1	95.2947
Ω_2	123.8832
ξ_1	0.05
ξ_2	0.0384

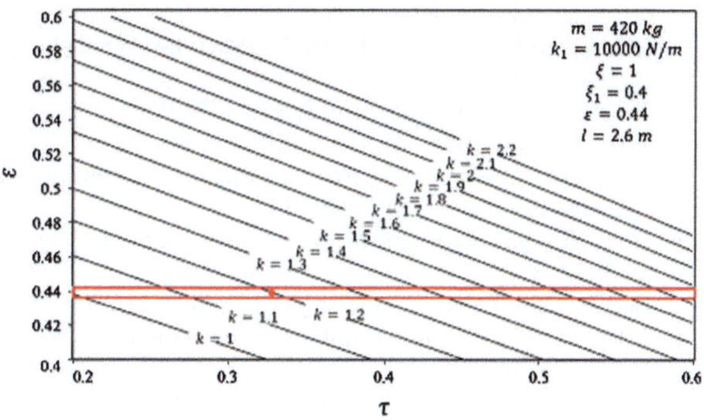

Fig. 10 Design chart for a smart suspension with a non-linear damper [15]

7 Conclusion

Olley's flat ride tuning has been regarded as a rule for designing chassis. The fact that these rules were based on experimental results, motivated many researchers to study and validate these rules. In this study, has been tried to validate Olley's results analytically for the first time.

As a result of the dual behavior of the suspension which is required to get the optimal flat ride, more accurate results were looked for using a nonlinear suspension system for this analysis. The results prove that the forward speed of the vehicle affects the flat ride condition, which agrees with previous researchers' results. In a passive suspension system flat ride can be achieved at a certain speed only, so the suspension system of a car should be designed in a way which provides the flat ride at a certain forward speed [16–19].

A design chart based on the nonlinear analysis, for smart active suspension systems has been provided which enables a car with smart suspension system to provide flat ride at any forward speed of the vehicle. The design chart can be used for designing chassis with passive suspension for a specified speed as well. Examples of both of the above mentioned conditions have been reviewed and discussed, by using some numerical values from a sample car. The research proves the effectiveness of Olley's flat ride for getting a more comfortable ride in cars, and considering the shortcomings of the principles suggests better ways of implementing them to the design of suspension systems for a better and more effective flat ride tuning.

References

1. Milliken WF, Milliken DL, Olley M (2002) Chassis design. Professional Engineering Publishing, London
2. Azizan MA, Fard M, Azari MF (2014) Characterization of the effects of vibration on seated driver alertness. Nonlinear Eng 3(3):163–168
3. Helmkamp JC, Talbott EO, Marsh GM (1984) Whole body vibration-a critical review. Am Ind Hyg Assoc J 45(3):162–167
4. Tanaka M, Mizuno K, Tajima S, Sasabe T, Watanabe Y (2009) Central nervous system fatigue alters autonomic nerve activity. Life Sci 84(7):235–239
5. Griffin MJ (1976) Subjective equivalence of sinusoidal and random wholebody vibration. J Acoust Soc Am 60(5):1140–1145
6. Olley M (1934) Independent wheel suspension—its whys and wherefores. Soc Automot Eng J 34(3):73–81
7. Best A (2002) Vehicle ride-stages in comprehension. Phys Technol 15(4):205
8. Olley M (1938) National influences on American passenger car design. Proc Inst Automob Eng 32(2):509–572
9. Olley M (1946) Road manners of the modern car. Proc Inst Automob Eng 41(1):147–182
10. Rowell HS, Guest JJ (1923) Proc Inst Automob Eng 18:455
11. Crolla D, King R (2000) Olley's 'Flat Ride' revisited
12. Sharp R (2002) Wheelbase filtering and automobile suspension tuning for minimizing motions in pitch. Proc Inst Mech Eng Part D: J Automob Eng 216(12):933–946
13. Sharp R, Pilbeam C (1993) Achievability and value of passive suspension design for minimum pitch response. Veh Ride Handl 39:243–259
14. Odhams A, Cebon D (2006) An analysis of ride coupling in automobile suspensions. Proc Inst Mech Eng Part D: J Automob Eng 220(8):1041–1061
15. Dai L, Jazar RN (eds) Nonlinear approaches in engineering applications. Chapter 1: Smart flat ride tuning. Springer, New York. http://www.springer.com/materials/mechanics/book/978-1-4614-6876-9
16. Marzbani H, Jazar R, Fard M (2013) Hydraulic engine mounts: a survey. J Vib Control 19(16)
17. Marzbani H, Jazar RN, Khazaei A (2012) Smart passive vibration isolation: requirements and unsolved problems. J Appl Nonlinear Dyn 1(4):341–386
18. Marzbani H, Jazar RN (2014) Smart flat ride tuning. Nonlinear approaches in engineering applications 2. Springer, New York, pp 3–36
19. Marzbani H (2012) Flat ride; problems and solutions in vehicle. Nonlinear Eng 1 (3–4):101–108

Part III
Engine Technologies

Gasoline Combustion System Development for Volvo Cars All-New Engine Family

Roy Ogink

Abstract Volvo Car Corporation has recently launched the first three members of a completely new engine family called Drive-E. It consists of four gasoline and four diesel engines that are characterized by high specific performance, low friction and low fuel consumption. The number of cylinders is limited to four, which implies that these down-sized, boosted engines will replace present five and six cylinder engines. This paper focuses on the development of the Drive-E gasoline combustion system, characterized by a centrally-mounted direct-injection fuel injector in combination with high-tumbling intake ports. The break-down of the large-scale tumble motion into a high level of turbulence leads to fast and stable combustion as well as excellent air/fuel mixing and limited wall wetting. The four gasoline engines span a wide range of power levels. This means that the engine at the entry power level needs to breathe less air than the 2.0 L engine at the highest power level (225 kW, 306 HP). Consequently, intake port design was an important part of the combustion system development. Different intake port geometries are applied at the different power levels, and this paper describes the trade-off between a design that sets up a strong tumble motion (leading to fast combustion and thus reduced fuel consumption), and a design that optimizes engine breathing (enabling class-leading specific power, while keeping fuel-efficiency at a good level).

1 Introduction

A few years ago, Volvo Car Corporation started the development of a completely new modular engine family called Drive-E. At that moment a wide range of four, five and six cylinder engines had to be fitted into Volvo's car models, giving rise to a large number of engine installations. The aim of the new engine program was to significantly reduce complexity and fuel consumption and it was decided to focus

R. Ogink (✉)
Volvo Car Corporation, Gothenburg, Sweden
e-mail: Roy.Ogink@volvocars.com

I. Denbratt et al. (eds.), *Sustainable Automotive Technologies 2014*,
Lecture Notes in Mobility, DOI 10.1007/978-3-319-17999-5_5

Fig. 1 CFD streamlines indicating the tumble motion set up by the intake ports

on four cylinder engines only. By means of down-sizing to a displacement of 2.0 L and applying different boosting systems, the Drive-E engines will replace all five and six cylinder engines in Volvo Cars' present line-up.

This paper describes the development process of the gasoline combustion system for the Drive-E family. The system relies heavily on a high-tumble concept (See Fig. 1) [1]. Since filling is less of an issue for turbo-charged engines than it traditionally has been for naturally-aspirated engines, it becomes possible to design intake ports that set-up a strong tumbling motion inside the combustion chamber during the period the intake valves are open. The large-scale tumble motion will survive during most of the compression stroke, after which it has to be broken down into small-scale turbulence. A properly designed combustion system will then be characterized by a high level of turbulent kinetic energy inside the combustion chamber prior to, and during, combustion. Consequently, the combustion process will be stable and fast. The following typical advantages can be mentioned for a high-tumble system:

- Shortened duration from spark until combustion start, as well as fast and stable combustion.
- Increased knock resistance, implying that a spark timing closer to optimal ignition timing can be used.
- Better fuel consumption at part-load conditions due to improved capability to handle internal EGR.
- More stable combustion at low engine speeds resulting in higher BMEP, since the engine can cope with higher boost pressures.
- Higher power levels can be reached at high engine speeds until reaching the maximum exhaust temperature and/or fuel enrichment level of the engine.
- The large-scale tumble motion, followed by the high level of turbulence, will enhance air/fuel mixing.

Faster combustion at boosted conditions might lead to an increased cylinder pressure rise rate, which indicates one disadvantage of high-tumble combustion systems: Too fast combustion might lead to NVH problems.

Table 1 Main properties for the 2.0 L, 4-cylinder T4, T5 and T6 Drive-E engines [2]

Engine	T6	T5	T4
Rated power (kW/hp)	225/306	180/245	140/190
Torque/ overboost (Nm)	400	350/380	300/330
Compression ratio (-)	10.3 : 1	10.8 : 1	11.3 : 1
Bore/stroke (mm)	82/93.2	82/93.2	82/93.2
Charging system	One roots + one turbo with waste gate	One turbo with waste gate	One turbo with waste gate

The gasoline engines in the Drive-E engine family will be available at different power levels (See Table 1) The two variants with the best performance (T5 and T6) are already available to customers, while other engines having slightly reduced performance but further improved fuel consumption (e.g. T4) will be released in the near future. The engines have many properties in common, such as bore, stroke and displaced volume. On the other hand, the engine with the highest performance, the T6, utilizes an advanced boosting system that consists of both a turbocharger and a supercharger, whereas the other engines are equipped with more conventional charging systems. The compression ratio is reduced in steps as the specific performance increases from T4 to T5 to T6. As far as the combustion system is concerned, the T5 and T6 engines share similar intake ports, while the lower performance variants have intake ports that focus even more on setting up a strong tumble motion than on filling the combustion chamber with fresh charge.

2 Gasoline Combustion System Development Process

Intake port design is the most critical step of the Drive-E gasoline combustion system development, since it defines the trade-off between tumble level and filling. As soon as the intake port geometry has been decided, the so-called spray targeting can be investigated, which defines the directions for the 6 spray plumes of the centrally-mounted injector in the Drive-E cylinder head. Since the strong air jet set up by the intake ports can significantly affect the fuel spray, the design of the intake ports and the choice of spray targeting are connected to each other to a very high degree. Together, the intake ports and the spray targeting are dominating the main characteristics that the combustion systems will have. Intake port design for Drive-E is discussed in detail in this paper. Spray targeting, and design of other parts of the combustion system (such as exhaust ports, piston and flame propagation) are not described here.

3 Intake Port Design

The design process for new intake ports can be divided into several phases. During the first phase, a set of intake port designs is evaluated by means of steady-state CFD. It is a very powerful tool to quickly compare the mass flow and tumble levels for a large number of intake port geometries. The results can be added to a scatter plot (See Fig. 2) to compare the mass flows and tumble levels for new intake port designs to each other, and to have a reference with engines developed in the past. For a certain valve lift, the scatter plot shows a very linear trade-off between filling and tumble level. Consequently, the aim is to find an intake port design that provides enough mass flow to reach the requested power level for the particular engine, and that has a tumble ratio that is as high as possible.

Besides results for mass flow and tumble, the velocity vectors of the gas flow in the CFD geometry are evaluated to understand how the present port design can be improved. Based on the flow (separation) in the ports and around the valves, input is provides for the next CFD-iteration to improve mass flow and/or tumble.

The outcome of the steady-state CFD analyses might e.g. be three promising designs, which are then further evaluated in the second phase: transient in-cylinder CFD. In these transient CFD analyses, different intake port designs are ranked for several operating conditions based on the following criteria: (1) filling (trapped mass prior to combustion), (2) air/fuel mixing, (3) the tumble level, (4) the level of turbulent kinetic energy (TKE) in the chamber prior to combustion, and (5) wall wetting (droplet impact and liquid film on liner, piston, head and valves).

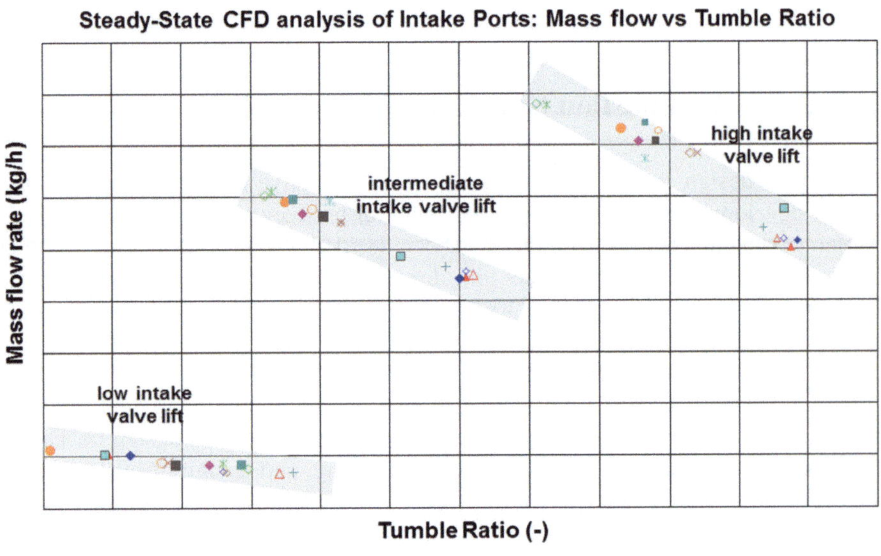

Fig. 2 Example of scatter plot obtained from steady-state intake port CFD

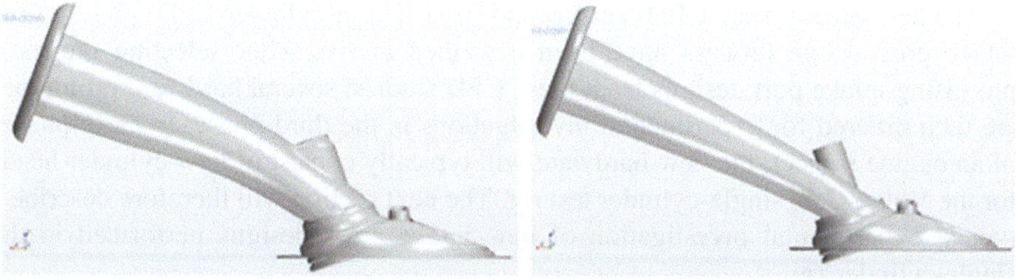

Fig. 3 Intake port design for the Fill port (*left*) and the Tumble port (*right*)

Below, CFD results for two intake port designs are discussed. For simplicity, these will be called Fill port (corresponding to the intake port for T5 and T6) and Tumble port (for the T4 power level) in the discussion below. The Fill port has a general shape that is more bend (See Fig. 3), which is generally better for achieving a higher mass flow. The flow will enter the combustion chamber in a more vertical way (compared with the Tumble port) and consequently the circumferential flow area around the valves can be better utilized by the flow. The Tumble port, on the other hand, has a much more straight design. Consequently, most of the incoming air is forced to enter the combustion chamber at the upper side of the intake valves.

The transient CFD runs have been performed for maximum power conditions (for the T5-position). As can be seen in Fig. 4, the tumble level of the Fill port is much lower during the intake and compression strokes than the level for the Tumble port. Since the large-scale tumble motion is broken down into turbulence, the average level of TKE at the end of the compression stroke is much higher for the Tumble port than for the Fill port (See Fig. 4). It should however be noted that the average TKE level quickly decays after 690 CAD. Furthermore, the transient CFD results showed: (1) a higher trapped mass after IVC for the Fill port than for the Tumble port design, and (2) further improved air/fuel mixing for the Tumble port compared with the Fill port.

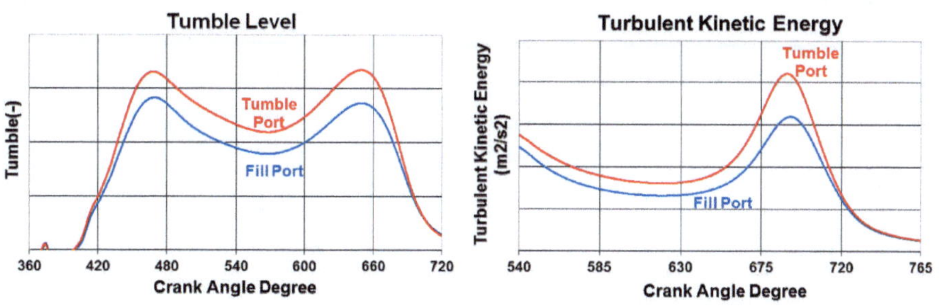

Fig. 4 Transient CFD results for Tumble and Fill ports: turbulent Kinetic Energy (*right*) and Tumble level (*left*)

The first (steady-state CFD) and second (transient in-cylinder CFD) phase of the intake port design process have been described above. After selecting the most promising intake port designs from these CFD studies, several hardware prototypes are then ordered for experimental investigations in the third phase. In early phases of an engine project, the new hardware will typically consist of new cylinder heads for the Volvo Cars single-cylinder test rig. The next section will therefore describe a typical experimental investigation of new intake ports designs performed in the single-cylinder rig.

3.1 Single-Cylinder Testing for Intake Port Design Evaluation

To analyze the behavior of new intake port designs, a cylinder head with the new port geometry is mounted on a single-cylinder engine. Two types of investigations are then performed to discriminate between port designs: (1) a so-called mini-map investigation consisting of 14 part-load operating points representative for the most common driving cycles and/or real customer driving. (2) a so-called Wide-Open-Throttle (WOT) investigation to study the engine's behavior at high load. The boundary conditions for the single-cylinder experiments were adopted from multi-cylinder experiments at the T5-power position. These boundary conditions are: boost pressure, exhaust back pressure, lambda, charge air temperature and fuel pressure. This section will discuss mini-map and WOT results for the Fill and the Tumble intake port designs.

The results from the mini-map investigation, which provides a tool to investigate part-load fuel consumption, can be summarized as follows (results are average values for the 14 mini-map points):

- The Tumble port has 1.5 % lower specific fuel consumption than the Fill port. Partially, this is due to the slightly higher compression ratio of the 1-cylinder with the Tumble-port (10.82 : 1) than the one of the Fill-port (10.75 : 1). (See Fig. 5)

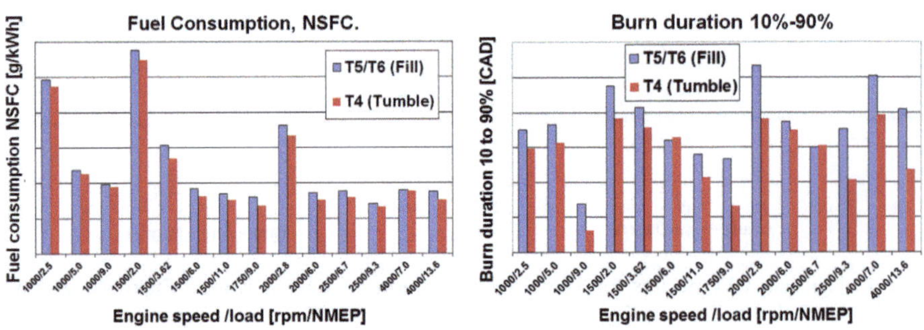

Fig. 5 Mini-map results from 1-cylinder testing: burn duration (*right*) and NSFC (left)

- The average burn duration (10–90 % of the fuel burnt) of the Tumble-port is 4 crank angle degrees (CAD) shorter (See Fig. 5).
- The Tumble-port has 0.9 % (absolute) higher combustion efficiency.
- The emissions of CO_2 are 0.16 % (absolute) higher for the Tumble-port, indicating better air/fuel mixing for this port: As expected, a higher level of large-scale motion and the consequent increased level of in-cylinder turbulence, contribute to mixing.
- Both the Fill- and Tumble-ports show low smoke emissions, indicating good mixing, but the Tumble-port has levels that are lower or very much lower than the results for the Fill-port.
- The Tumble-port has 21 % lower emissions of unburnt hydrocarbons (HC).

These mini-map investigations clearly show the benefits obtained with the Tumble-port design during part-load operation.

As described above, WOT-testing is the second way to experimentally compare intake port designs. Such tests have been performed from 1500 to 5700 RPM to compare the WOT-characteristics of the Tumble- and Fill-ports. For each operating point, the load is increased until knock reaches a pre-defined level. In this way, the maximum achievable torque is obtained for the whole engine speed range. Different intake port designs will have different properties affecting the knock resistance of the engine design, such as in-cylinder turbulence levels (affecting combustion duration) and air/fuel mixing.

As far as the results are concerned, it can be observed that the Tumble-port gives rise to: (1) a 2–5 CAD faster combustion duration (See Fig. 6), (2) a slightly more advanced spark timing and (3) an increased cylinder pressure rise rate. The disadvantage of this tumble-optimized port is that filling is sacrificed above 3900 RPM (for these experiments with T5-power-level boundary conditions): for high engine speeds, the maximum achievable torque is significantly lower for the Tumble-port than for the Fill-port (See Fig. 6).

So, both the mini-map and WOT-investigations show significantly improved combustion properties when increasing the tumble level. One might argue that

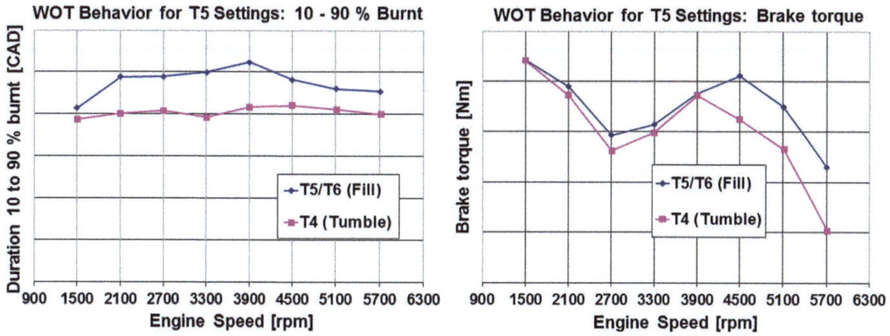

Fig. 6 WOT results from 1-cylinder testing: brake Torque (*right*) and 10–90 % Burn Duration (*left*)

increased tumble (and TKE) will lead to increased heat losses. Previous work by the authors indicated that tumble can affect heat losses, but that the shape of the air jet set up by the tumble motion can be just as important as the tumble level itself [1]. In the previous work it was also concluded that the positive effects of increased tumble (e.g. on efficiency) outweigh the eventual disadvantage of increased heat losses.

4 Conclusions

The development process of the gasoline combustion system for Volvo Cars new Drive-E engine family has been discussed with the focus on intake port design. Drive-E consists of four-cylinder engines characterized by high specific performance and low fuel consumption. By comparing a high-tumble intake port to a port more oriented towards filling, clear advantages were observed for the tumble port. Experimental investigations showed improved fuel consumption, better mixing, lower emissions (HC, CO and soot) and superior combustion properties for the tumble port design.

Acknowledgments The author would like to thank Roy Ekdahl, who performed the 1-cylinder experiments and Aristotelis Babajimopoulos for carrying out the CFD analyses. Many thanks also go to Göran Josefsson, who heavily contributed to the gasoline combustion system development for Drive-E by means of his broad and deep knowledge of combustion systems.

References

1. Berntsson A, Josefsson G, Ekdahl R, Ogink R, Grandin B (2011) The effect of tumble flow on efficiency for a direct injected turbocharged downsized gasoline engine. SAE paper 2011-24-0054
2. Crabb D, Fleiss M, Larsson J-E, Somhorst J (2013) New modular engine platform from Volvo. MTZworldwide 74(9):4

Use of Late IVC and EGR to Enhance Diesel Engine Optimization

Jonas Sjöblom

Abstract The increasing demand for improved efficiency of diesel engines requires more advanced combustion solutions. In addition to traditional methods such as EGR, turbocharging and advanced injection systems, variable valve timing is now available at a reasonable production cost. The use of variable inlet valve timing provides an efficient way for Low-Temperature Combustion (LTC) which provides high thermal efficiency in combination of low emission levels. Furthermore, by modifying the characteristics of the charge air (i.e. by means of EGR, boost pressure and late inlet valve closing, LIVC), further hardware optimization becomes possible, e.g. by increasing compression ratio without reaching critical peak pressures. In the present study, the effect of LIVC was investigated together with the effect of EGR in a single cylinder heavy duty diesel engine. The engine was equipped with pneumatically controlled inlet valves and a high pressure common rail injector. Different injection timings and injection pressures were investigated at two different load points. The mass flow of oxygen was kept constant in order to show how the physical properties (density and temperature) affect the combustion and emission characteristics. The combustion results showed that if the oxygen mass flow is kept constant, EGR is a more efficient way (compared to LIVC) to lower the fuel consumption since it is accompanied with the largest gas flow and thus increased fuel conversion efficiency. The LIVC decreased the fuel consumption at low loads and reduced the emissions at both loads. Transportation of people and goods tend to increase and since internal combustion engines will remain a major power supply for many years to come, reduced fuel consumption is an utmost important way to decrease the CO_2 emissions and to move towards a sustainable society. The results in this study show that variable inlet valve timing can be used as one important complementary tool to obtain better combustion characteristics and thus enabling more efficient powertrains.

J. Sjöblom (✉)
Applied Mechanics, Chalmers University of Technology, 41296 Gothenburg, Sweden
e-mail: jonas.sjoblom@chalmers.se

I. Denbratt et al. (eds.), *Sustainable Automotive Technologies 2014*,
Lecture Notes in Mobility, DOI 10.1007/978-3-319-17999-5_6

Abbreviations

atdc/abdc	After/Before top dead center
CA50	50 % of total heat release
CAD	Crank Angle Degrees
EGR	Exhaust Gas Recirculation
EOI	End of Injection
IMEP	Indicated mean effective pressure
ISFC	Indicated specific fuel consumption
IVC, IVC_{eff}	Inlet Valve Close, effective Inlet Valve Close
LIVC	Late Inlet Valve closing
LTC	Low-Temperature Combustion
P_{inj}	(rail) Injection Pressure
PPCI	Partially Pre-mixed Compression Ignition
RoHR	Rate of Heat Release
SOC	Start of Combustion
SOI	Start of Injection
tdc, bdc	Top dead centre, bottom dead centre
VVA	Variable Valve Actuation

1 Introduction

The world faces grand challenges to mitigate the effects of an un-sustainable usage of fossil fuels. While alternative fuels will become more important, fossil fuel will be a major energy source for transportation in many years to come. Improved efficiency of the internal combustion engines provides one important way to reduce CO_2 emissions and present and future legislations will make an important contribution to reduce these CO_2 emissions.

There are many ways to improve the efficiency but improved combustion efficiency is often accompanied with high local temperature which results in unacceptably high NO_X emissions. NO_X is mainly formed by the Zeldovic mechanism which is favored by high temperature (as well as oxygen concentration and time). One method to reduce NO_X formation is by lowering the oxygen concentration by means of EGR. The lowered oxygen concentration will also result in lower peak temperature and significant reductions in NO_X formation can be obtained. However, EGR also slows down the combustion rate and the efficiency is not as high as for non-EGR systems. Another way to obtain lower peak temperatures is by obtaining a cooler gas mixture at the time for fuel injection. Even though charge air cooling is commonly applied, the gas compression itself inevitably rises the temperature. Yet another NO_X reduction strategy is reducing the compression stroke itself by means of changing the timing of the inlet valve closure (IVC).

This method was proposed by Ralph Miller in 1957 [1] and it appears in the literature under a variety of names depending on the technical implementation. In this study we will use the name "late Miller" even though "supercharged Atkinson cycle" is also appropriate. Since reduced compression stroke will result in lower amount of air, the volumetric efficiency will be low unless compensated for by a turbocharging system. Until recently, systems for changing the IVC haven't been economically attractive for implementation in heavy duty powertrains, but upcoming regulation changes and the need for more fuel-efficient powertrains have changed this situation. In a current EU project (CORE, CO_2 Reduction for long distance transport) [2], different Variable Valve Actuation (VVA) systems are being evaluated and promising results are being obtained by coupling the approach with an efficient turbocharging system.

Over-expanded cycles are well established [3] showing an increase in efficiency but with a decrease in IMEP and power density. The Miller cycle has been studied for the Otto-cycle (e.g. [4, 5]), the diesel cycle (e.g. [6–14]) or both (e.g. [15, 16]). The concept has also been studied for alternative fuels (e.g. [17]). Supercharging was already envisaged in the Miller patent, but in cases where fuel efficiency alone is sought the Miller cycle without supercharging (i.e. the Atkinson cycle) is applied, for example in hybrid vehicles [18].

In the literature, different aspects of various Miller cycles have been investigated. Since the lower compression ratio lowers the peak temperature, many studies demonstrate reduced NO_X emissions e.g. [6, 7, 17, 19]. Another aspect of the system to consider is the need for supercharging during high load [11, 12]. The lower cylinder pressure (i.e. lower gas density) will increase the risk of the spray hitting the wall e.g. [6, 8, 20] which could be detected by particulate matter (PM) analysis [13]. Even though many studies report on the importance of spray atomization, most studies keep the injection pressure constant (e.g. [6, 13, 21–24]. In some cases the injection pressure is varied as a means to vary the load point [7, 17, 19].

Many studies explore the possibility of having a larger amount of pre-mixed combustion (e.g. [6, 24]) and the effects of EGR to lower both NO_X and peak pressures [5, 9, 10, 12].

In this study the effect of LIVC was investigated together with the effect of EGR in a single cylinder heavy duty diesel engine. Different injection timings and injection pressures were investigated at two different load points. The mass flow of oxygen was kept constant in order to show how the physical properties (density and temperature) affect the combustion and emission characteristics.

2 Experimental

The engine was a single cylinder heavy duty diesel engine equipped with a common rail injector system (from Delphi) and a VVA system (from Cargine). The EGR was controlled by adjusting the back pressure and in the case of no EGR the back pressure was set equal to the boost pressure, as in [9]. Ultra low sulfur diesel fuel

Table 1 Single cylinder engine and nozzle specifications

Displacement	2 dm^3
Bore/stroke	131/150 mm
Geometrical compression ratio	17:1
Nozzle design	5 hole, 150° angle
Nozzle flow number	2.00, d_0 = 256 μm

Table 2 Evaluated experimental conditions

Variable	Levels
Load points	1000 rpm/120 Nm, 1500 rpm/180 Nm
IVC (set values)	0; 40; 70; (90, 110) CADabdc
EGR	0; 20; 40 %
P_{inj}	1500; 2400 bar
CA50	Approx. 2; 5; 8 CADatdc

(Swedish MK1) was used throughout this study. The engine details are given in Table 1. The common rail system was pressurized with a pumping injector and can deliver rail pressures up to 3000 bar [25]. The rail pressure, injection timing and duration was controlled by a separate software (ATI Vision).

Two different load points were investigated; one low load (1000 rpm, 120 Nm, 25 % of full load) and one medium load (1500 rpm, 180 Nm, 50 % of full load). In order to isolate the physical properties of the charge air (temperature and density) from chemical property (oxygen concentration), the oxygen flow into the cylinder was kept constant by adjusting the Boost pressure. Other factors affecting the system were explored (by applying factorial designs) over the design space by varying IVC, EGR, injection pressure and timing, see Table 2. The start of injection (SOI) was adjusted to get 50 % burnt (CA50) at three different timings.

In total, about 120 experiments were performed and a number of intermediate responses were measured along with the end performance responses which were fuel consumption (ISFC) and emission (NO_X, CO, HC and soot) results.

Emissions (NO_X, CO and HC) were measured with an AVL AMAi60 and the soot was measured with a smoke meter (AVL 415). The rate of heat release was calculated in Osiris (D2T). The fuel consumption is given as ISFC (indicated specific fuel consumption) since compressor work is not included in the system but rail pressure is included (rocker arm driven by the camshaft).

3 Results and Discussions

3.1 Analysis of Pressures and Heat Release

A few examples of Rate of Heat Release (RoHR) are given in Fig. 1 (low load, without EGR) and Fig. 2 (high load, without EGR).

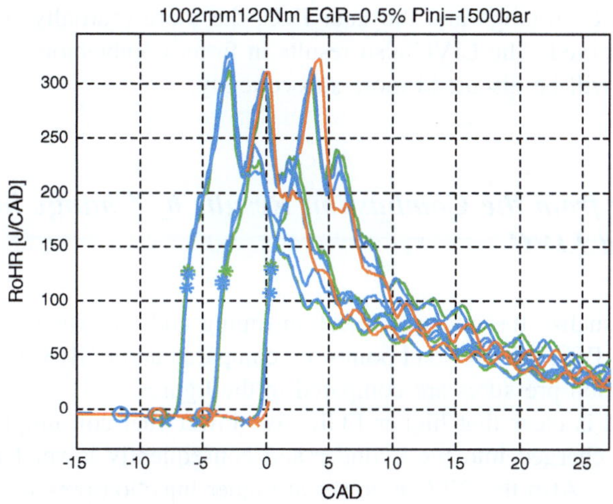

Fig. 1 Rate of heat release for experiments at low load and without EGR. *Circles* indicate SOI, *cross* indicates SOC and *stars* indicates EOI. Three different timings are displayed (CA50 = 2, 5 and 9 CADatdc). The *colors* represent different IVC (*green* $IVC_{set} = 0$, *blue* $IVC_{set} = 40$ CADabdc, *orange* $IVC_{set} = 70$ CADabdc)

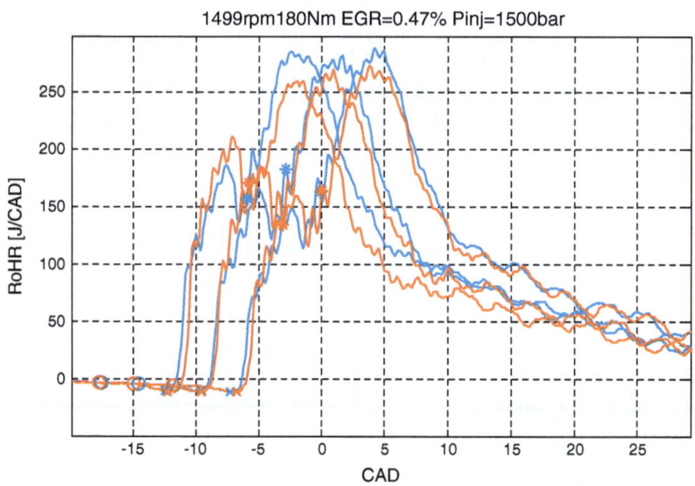

Fig. 2 Rate of heat release for experiments at high load and without EGR. *Circles* indicate SOI, *cross* indicates SOC and *stars* indicates EOI. Three different timings are displayed (CA50 = 1, 4 and 8 CADatdc). The *colors* represent different IVC (*blue* $IVC_{set} = 40$ CADabdc, *orange* $IVC_{set} = 70$ CADabdc)

The experiments at low load (Fig. 1) show more premixed combustion (the first peak of the RoHR is relatively larger than the later part) compared to the experiments at high load (Fig. 2). Generally (shown in both figures) the mode advanced (earlier) injection timings and the later IVC, the more (partially) premixed combustion is observed. The LIVC also results in faster combustion (narrower RoHR curve) which will be discussed more subsequently.

3.2 Results from the Combustion Results by Change in SOI, P_{inj} and LIVC

In order to visualize the effect of injection timing and the interaction of injection pressure and LIVC, the two load points are compared and EGR versus non-EGR at different injection pressures are compared in the figures.

In Fig. 3 it is clear that higher EGR give lower fuel consumption due to the increased gas charged into the cylinder and consequently lower temperature and less heat losses. Also the ISFC is lower at higher injection pressure (due to better fuel-air mixing and more efficient combustion). The changes in ISFC when varying LIVC are relatively small. However, from Fig. 3 it is clear that ISFC does not always decrease with increasing LIVC, e.g. at high load. The reason for different sensitivity on LIVC is because the combustion is different: In the low load case, the combustion is mainly partially pre-mixed combustion which benefices from LIVC whereas in the high load case, the combustion has more diffusion-controlled combustion and the lower temperature from LIVC is not beneficial for the ISFC.

The NO_X emissions are displayed in Fig. 4. The NO_X emissions are lower with EGR (due to the lower oxygen concentration) and NO_X emissions are also decreased by later injection (SOI) and later IVC. High injection pressure makes a

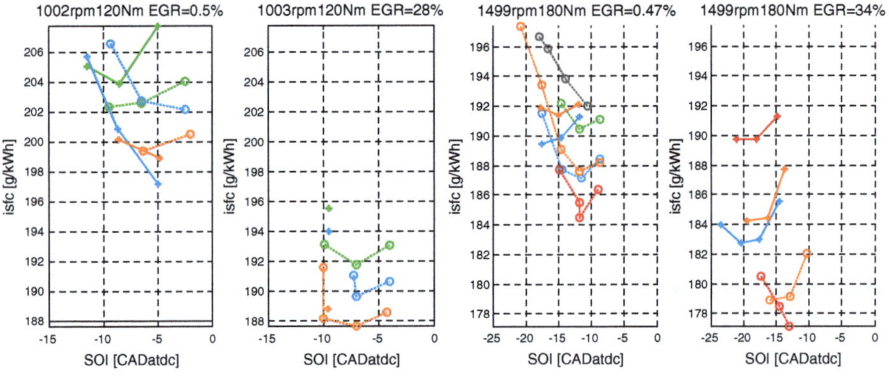

Fig. 3 Indicated fuel consumption versus SOI for two load points and different amount of LIVC, EGR and P_{inj}. Different injection pressures are indicated by *symbols* (*solid-cross* P_{inj} = 1500 bar, *dashed-circle* P_{inj} = 2400 bar). The *colors* represent different IVC (*green* IVC_{set} = 0 CADabdc, *blue* IVC_{set} = 40 CADabdc, *orange* IVC_{set} = 70 CADabdc, *red* IVC_{set} = 90 CADabdc, *black* IVC_{set} = 110 CADabdc)

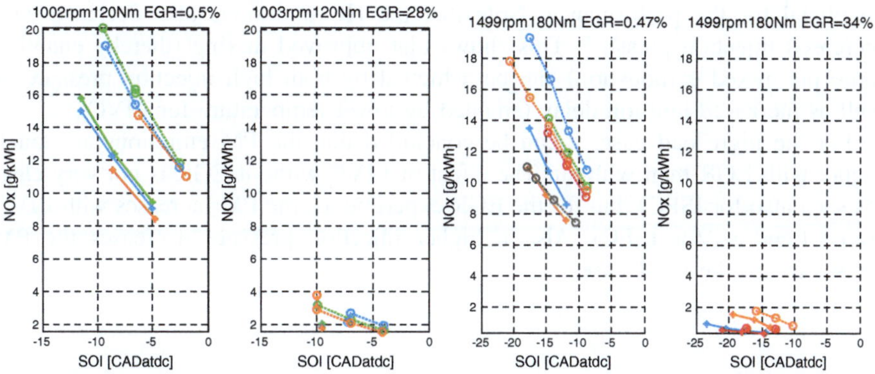

Fig. 4 NO_X emissions versus SOI for two load points and different amount of LIVC, EGR and P_{inj}. Different injection pressures are indicated by *symbols* (*solid-cross* P_{inj} = 1500 bar, *dashed-circle* P_{inj} = 2400 bar). The *colors* represent different IVC (*green* IVC_{set} = 0 CADabdc, *blue* IVC_{set} = 40 CADabdc, *orange* IVC_{set} = 70 CADabdc, *red* IVC_{set} = 90 CADabdc, *black* IVC_{set} = 110 CADabdc)

more rapid combustion and the NO_X emissions increases, especially for cases without EGR.

HC emissions (for the low load case) and PM emissions (for the high load case) are displayed in Fig. 5. The reason for displaying this combination is because the PM emissions for the low load were zero for all cases. The HC emissions for the high load shows similar trend as for the PM (increasing emissions with later SOI).

From Fig. 5 it is clear that the combustion processes are different for the two load points as already seen for the ISFC. For the low load point, the combustion improves with later IVC, since increased amount of pre-mixed combustion is

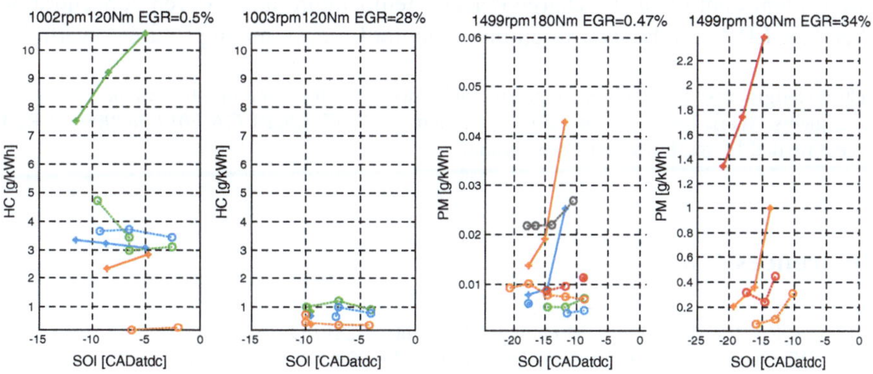

Fig. 5 HC emissions for the low load point (*left*) and PM emissions for the high load point (*right*) versus SOI and different amount of LIVC, EGR and P_{inj}. Different injection pressures are indicated by *symbols* (*solid-cross* P_{inj} = 1500 bar, *dashed-circle* P_{inj} = 2400 bar). The *colors* represent different IVC (*green* IVC_{set} = 0 CADabdc, *blue* IVC_{set} = 40 CADabdc, *orange* IVC_{set} = 70 CADabdc, *red* IVC_{set} = 90 CADabdc, *black* IVC_{set} = 110 CADabdc)

beneficial for the performance. Note that the HC emissions also decrease with increased injection pressure. This shows that improved mixing (thereby enabling more pre-mixed combustion) can be achieved by both high injection pressure as well as increased ignition delay (enabled by lower temperature for LIVC).

For the high load case, it can be concluded that the PM emissions are much higher with EGR than without. The effect of LIVC without EGR is not very clear (as seen also for ISFC), but for the EGR experiments, the PM increases with LIVC (same trend as for ISFC). Also a higher injection pressure decreases the PM emissions for the EGR case.

4 Conclusions

In the presented investigation, LIVC has been explored by varying most engine parameters except the oxygen flow. One consequence from this methodology is that higher EGR requires higher charge gas flow which gives better efficiency. From the experiments performed here, one might be tempted to conclude that EGR provides a more efficient means to reduce ISFC than LIVC, but the current experimental setup cannot prove this conclusion since the EGR is highly correlated with the gas flow. This issue is presently under investigation.

The sensitivity of the results for LIVC at different load points are somewhat different which indicates that the two load point have different types of combustion where the high load had more of a traditional Diesel combustion. Therefore, general statements of the benefits of LIVC cannot be given without characterizing the combustion conditions (such as injection pressure, Boost pressure and EGR). However, the results show that the combustion characteristics are very sensitive to LIVC. In combination with turbo charging and EGR, the use of LIVC enables independent control of the charge density, temperature and oxygen concentration. Therefore LIVC can be used as an important tool for optimizing future powertrains.

Acknowledgments Savo Girja is greatly acknowledged for running the engine during the experiments. Work has been carried out as a part of CORE (SCP1-GA-2012-284909), a R&D project funded by the European Commission.

References

1. Miller R (1957) Supercharged engine. US patent US2817322 A
2. CORE (2011) EU-project, GA SCP1-GA-2012-284909. www.co2re.eu
3. Heywood JB (1988) Chapter 5: Ideal models of engine cycles. In: Internal combustion engine fundamentals. McGraw-Hill, Singapore. ISBN 0-07-100499-9
4. Ebrahimi R (2012) Performance analysis of an irreversible Miller cycle with considerations of relative air–fuel ratio and stroke length. Appl Math Model 36(9):4073–4079. doi:10.1016/j.apm.2011.11.031

5. Hatamura K, Hayakawa M, Goto T, Hitomi M (1997) A study of the improvement effect of Miller-cycle on mean effective pressure limit for high-pressure supercharged gasoline engines. JSAE Rev 18(2):101–106. doi:10.1016/S0389-4304(96)00069-0
6. Murata Y, Nishio Y, Kusaka J, Daisho Y et al (2010) Numerical analysis of Miller-premixed charge compression ignition combustion on a dynamic phi-T map. Int J Engine Res 11(2):89–98
7. Imperato M, Antila E, Sarjovaara T, Kaario O et al (2009) NOx reduction in a medium-speed single-cylinder diesel engine using Miller cycle with very advanced valve timing. SAE technical paper 2009-24-0112 2009-24-0112. doi:10.4271/2009-24-0112
8. Brijesh P, Sreedhara S (2013) Exhaust emissions and its control methods in compression ignition engines: a review. Int J Automot Technol 14(2):195–206. doi:10.1007/s12239-013-0022-2
9. Ehleskog M, Gjirja S, Denbratt I (2012) Effects of variable inlet valve timing and swirl ratio on combustion and emissions in a heavy duty diesel engine. SAE technical paper 2012-01-1719. doi:10.4271/2012-01-1719
10. Edwards SP, Frankle GR, Wirbeleit F, Raab A (1998) The potential of a combined Miller cycle and internal EGR engine for future heavy duty truck applications. SAE technical paper 980180. doi:10.4271/980180
11. Millo F, Mallamo F, Mego GG (2005) The potential of dual stage turbocharging and Miller cycle for HD diesel engines. SAE technical paper 2005-01-0221. doi:10.4271/2005-01-0221
12. Millo F, Bernardi MG, Delneri D (2011) Computational analysis of internal and external EGR strategies combined with Miller cycle concept for a two stage turbocharged medium speed marine diesel engine. SAE Int J Engines 4(1):1319–1330. doi:10.4271/2011-01-1142
13. Benajes J, Novella R, Arthozoul S, Kolodziej C (2010) Particle size distribution measurements from early to late injection timing low temperature combustion in a heavy duty diesel engine. SAE Int J Fuels Lubr 3(1):567–581. doi:10.4271/2010-01-1121
14. Benajes J, Molina S, Martín J, Novella R (2009) Effect of advancing the closing angle of the intake valves on diffusion-controlled combustion in a HD diesel engine. Appl Therm Eng 29 (10):1947–1954. doi:10.1016/j.applthermaleng.2008.09.014
15. Martins JJG, Uzuneanu K, Ribeiro BS, Jasasky O (2004) Thermodynamic analysis of an over-expanded engine. SAE technical paper 2004-01-0617. doi:10.4271/2004-01-0617
16. Ribeiro B, Martins J (2007) Direct comparison of an engine working under Otto, Miller and diesel cycles: thermodynamic analysis and real engine performance. SAE technical paper 2007-01-0261. doi:10.4271/2007-01-0261
17. Heikkilä J, Happonen M, Murtonen T, Lehto K et al (2012) Study of Miller timing on exhaust emissions of a hydrotreated vegetable oil (HVO)-fueled diesel engine. J Air Waste Manag Assoc 62(11):1305–1312. doi:10.1080/10962247.2012.708383
18. Higuchi N, Sunaga Y, Tanaka M, Shimada H (2013) Development of a new two-motor plug-in hybrid system. SAE Int J Alt Power 2(1):135–145. doi:10.4271/2013-01-1476
19. He X, Durrett RP, Sun Z (2008) Late intake valve closing as an emissions control strategy at Tier 2 Bin 5 engine-out NOx level. SAE Int J Engines 1(1):427–443. doi:10.4271/2008-01-0637
20. Doosje E, Willems F, Baert R, Van Dijk M (2012) Experimental study into a hybrid PCCI/CI concept for next-generation heavy-duty diesel engines. SAE technical paper 2012-01-1114. doi:10.4271/2012-01-1114
21. De Ojeda W (2010) Effect of variable valve timing on diesel combustion characteristics. SAE technical paper 2010-01-1124. doi:10.4271/2010-01-1124
22. Wu B, Yu H, Pak P, Pei Y et al (2013) Effects of late intake valve closing timing on thermal efficiency and emissions based on a two-stage turbocharger diesel engine. SAE technical paper 2013-01-0276. doi:10.4271/2013-01-0276
23. Benajes J, Serrano JR, Molina S, Novella R (2009) Potential of Atkinson cycle combined with EGR for pollutant control in a HD diesel engine. Energy Convers Manag 50(1):174–183. doi:10.1016/j.enconman.2008.08.034

24. Murata Y, Kusaka J, Daisho Y, Kawano D et al (2008) Miller-PCCI combustion in an HSDI diesel engine with VVT. SAE Int J Engines 1(1):444–456. doi:10.4271/2008-01-0644
25. Delphi (2014) Delphi F2e ultra high pressure heavy duty diesel common rail system [cited 2014 2014-04-23; F2 common rail system]. http://delphi.com/manufacturers/cv/powertrain/common-rail-systems/f2-distributed-pump-diesel/

Numerical and Experimental Study of Stratified Turbulent Combustion in a Spray-Guided Gasoline Direct Injection Engine

Chen Huang, Andrei Lipatnikov, Lars Christian Riis Johansen and Stina Hemdal

Abstract Direct Injection (DI) of gasoline into cylinder of a Spark Ignition (SI) engine is widely recognized to be a promising technology capable for significantly reducing fuel consumption and carbon dioxide emissions as compared to a port-fuel injection SI engine. In particular, spray-guided (SG) GDI combustion systems allow for further improvement in fuel efficiency. Moreover, efficient CFD tools for numerical simulations of spray and combustion processes have been becoming increasingly important in engine development. In previous papers, a so-called Flame Speed Closure (FSC) model was implemented into an open source code OpenFOAM® with the capability of addressing important phenomena in SG GDI engines, e.g. fluctuations in mixture composition and the proper evaluation of combustion temperature for the products. In this paper, the aforementioned FSC model is applied to investigate the stratified turbulent combustion in a SG GDI engine in the frame work of unsteady 3D Reynolds-Averaged Navier–Stokes (RANS) simulations. The computed results are compared with the measured pressure traces obtained in the same research group for both low and medium load conditions. Further on, the calculated Reynolds-averaged progress variable is compared to the experimentally observed images.

1 Introduction

Limited oil resources and more stringent emission legislations especially for carbon dioxide (CO_2) are the driving forces to develop efficient and clean internal combustion engines (ICE) for passenger cars. Gasoline driven cars dominate the major markets such as US, China and Japan with diesels playing almost no role [1]. Even though the European market is traditionally dominated by diesels, there is a striking

C. Huang · A. Lipatnikov (✉) · L.C.R. Johansen · S. Hemdal
Department of Applied Mechanics, Chalmers University of Technology, Gothenburg, Sweden
e-mail: Andrei.Lipatnikov@chalmers.se

I. Denbratt et al. (eds.), *Sustainable Automotive Technologies 2014*, Lecture Notes in Mobility, DOI 10.1007/978-3-319-17999-5_7

trend that Gasoline Direct Injection (GDI) vehicles have been increased sharply since 2008 with an estimated market share of 25 % in 2012 [1]. GDI technology has a great potential of reducing CO_2 emission by 20–26 % for passenger cars as compared to port-fuel injected (PFI), spark-ignition, natural aspirated cars [2]. The SG GDI combustion system with a high pressure (around 200 bar) outward-opening piezo-controlled pintle-type injector represents the second generation of modern GDI engines; an optimized SG system has very high potential in fuel economy, around 4–6 % less than for wall-guided system (the first generation of GDI system) [3]. The significant reduction in CO_2 emission for SG GDI engine is due to (i) the reduction of pumping losses by removing the throttle, (ii) cooling of charge due to the evaporation of fuel spray, (iii) a higher compression ratio due to the reduction of knock propensity, (iv) reduced heat losses, etc. [4]. However, one of the challenges in developing such engines is the stringent requirement in positioning the spark plug in relation to the spray in order to ensure an ignitable fuel-air mixture with steep stratification gradient right between the ignition electrodes for a wide range of operating conditions.

Optical measurements and numerical simulations are important tools of R&D in the field of SG GDI engines. Moreover, experimental data especially that obtained with well-defined initial and boundary conditions is crucial for validation and development of numerical models. In particular, CFD simulations run by using an open source code OpenFOAM® has been attracting growing interests from both industries and academies due to the possibility to access the source code without paying license fees.

In previous papers, a so-called Flame Speed Closure (FSC) model for premixed turbulent combustion was combined with a presumed Probability Density Function (PDF) approach to address the effect of fluctuations in mixture composition on burning rate [5, 6]. Such combined model was implemented into OpenFOAM® to study the stratified turbulent combustion in a SG GDI engine.

Accordingly, this paper aims at (i) applying the combined model for simulating fuel injection and stratified turbulent combustion in a SG GDI engine and (ii) comparing the simulated results with measured data in terms of pressure traces and flame propagation images.

2 Method

The complex nature of reacting flows in SG GDI engines involves many physical and chemical phenomena which need to be included in unsteady 3D RANS modelling. The model used in this work includes many important phenomena, which can be categorized as follows: (i) spray (fuel injection, breakup, vaporization) and (ii) turbulent combustion (ignition, turbulent flame propagation in highly turbulent and inhomogeneous mixture). A brief discussion of such model is presented here; interested readers are referred to Refs. [5, 6, 7] for further details.

As far as the spray simulation is concerned, a uniform droplet size of 30 μm and the Kelvin-Helmholtz Rayleigh-Taylor (KHRT) model for secondary breakup were invoked. The spray-wall interaction model was a reflect model existing in the standard OpenFOAM®. Huang and Lipatnikov [7] simulated gasoline hollow-cone sprays and the calculated liquid penetration lengths and Sauter Mean Diameters (SMDs) were validated against measured data provided by Hemdal et al. [8] in Chalmers spray rig.

As far as the stratified turbulent burning is concerned, two important phenomena are addressed by FSC model combined with presumed PDF method. First, in order to account the effect of local fuel-air mixture on turbulent burning rate, accurate prediction of turbulent flame speed, thus, laminar flame speed under high temperature and pressure conditions are of crucial importance. Therefore, an approximation of the laminar flame speed of gasoline-air mixture as a function of equivalence ratio ϕ, pressure p and temperature T_u under conditions of $0.2 \leqslant \phi \leqslant 2.0$, $1 \le p \le 30$ atm, $298 \leqslant T_u \leqslant 800$K, (see Appendix B in Ref. [5]) was invoked in present unsteady 3D RANS simulations. This approximation was obtained by using a combustion chemistry featuring 120 species and 677 reactions for gasoline surrogate, i.e. a blend of *iso*-octane, toluene, and *n*-heptane in volumetric proportions of 55 : 35 : 10 %, respectively [5, 9, 10]. The second important phenomenon addressed by the combustion model is the effect of fluctuations in mixture fraction on turbulent burning rate. Such phenomenon was modelled by combining the FSC model for premixed turbulent combustion with a presumed PDF in a form of beta function [5, 6].

3 Experimental and Numerical Setups

Experiments were conducted on a single-cylinder research engine (AVL 5411.018) with optical access through a quartz window in the piston (Bowditch design) and the upper part of the cylinder liner. The engine was operated in stratified mode with wide open throttle at fixed rotational speed of 1500 rpm and at two load cases with target imeps of 3.8 and 5 bar. The engine specifications are listed in Table 1. The fuel used was gasoline 95 octane with 10 % ethanol. The fuel was injected directly into the cylinder using an outward-opening piezo-actuated injector and the resulting fuel cloud was ignited by a three-electrode spark plug. The injector and spark plug was mounted between the intake and exhaust valves in a close couple configuration, and they were positioned with an angle with respect to the vertical line of 20° and 5°, respectively. The fuel was injected at an injection pressure of 200 bar using a single injection strategy and the injection duration was 420 μs for the low load case and 550 μs for the medium load case. For the low load case the fuel was injected at 20 CAD bTDC and the spark timing was 16 CAD bTDC. For the medium load case the injection and sprak timing was 32 and 20 CAD bTDC, respectively. The in-cylinder pressure was measured using a pressure transducer for 100 firing cycles. The fuel injection, spark and following combustion were captured using high speed

Table 1 Engine specifications

Bore (mm)	83
Stroke (mm)	90
Compression ratio	10.2
Displacement volume (cm^3)	487
Dead volume (cm^3)	53
Intake valve diameter (mm)	33
IVO/IVC (CAD)	340/600
Exhaust valve diameter (mm)	28
EVO/EVC (CAD)	105/365

video with a frame rate of 9000 images per second. The images of the combustion were processed to achieve average images of seven individual cycles.

Because the focus of present paper is placed on the stratified turbulent combustion modelling and manipulation of moving valves in OpenFOAM® has been complicated and computationally demanding, the present simulations were performed for a closed cycle, i.e. from the Intake Valve Closing (IVC) until the Exhaust Valve Opening (EVO). A full cycle simulation including intake process was run by using KIVA code to estimate the initial conditions including turbulence kinetic energy (3.6 m^2/s^2) and its dissipation rate (1210 m^2/s^3). The initial tumble direction is shown in Fig. 1 with a maximum velocity of around 15 m/s. Fixed temperature boundary conditions were set at the piston, cylinder linear, and cylinder head, with the wall temperature in each boundary being 500, 480, and 450 K, respectively.

A computational mesh that features the combustion chamber geometry of the optical engine (see Fig. 1), was generated using ICEM CFD® and then converted into OpenFOAM®. This mesh features 62 756 cells with mesh size of around 1–2 mm in each direction. The time step used in this work was 0.1 CAD. Computations were run on AMD Opteron 6220 CPUs, on a single node with 16 cores and 32/64 GB of RAM. Depending on the complexity of the sub-models, the computational time typically ranged from 17 h to 3 days.

4 Results and Discussions

Comparisons of the computed and measured in-cylinder pressure curves for two different loads are shown in Fig. 2. It is worth noting that the pressure traces for each load were measured for around 100 cycles (black solid lines in Fig. 2) with the ensemble-averaged results shown in yellow solid lines with circles in Fig. 2. In general, the computed pressure traces agree quite well with the ensemble-averaged ones especially for the pressure rise rate during the combustion phase. For the low load case, i.e. imep = 3.8 bar, the computed pressure trace rises slightly earlier than that in the measurement. The main cause for such a difference is the

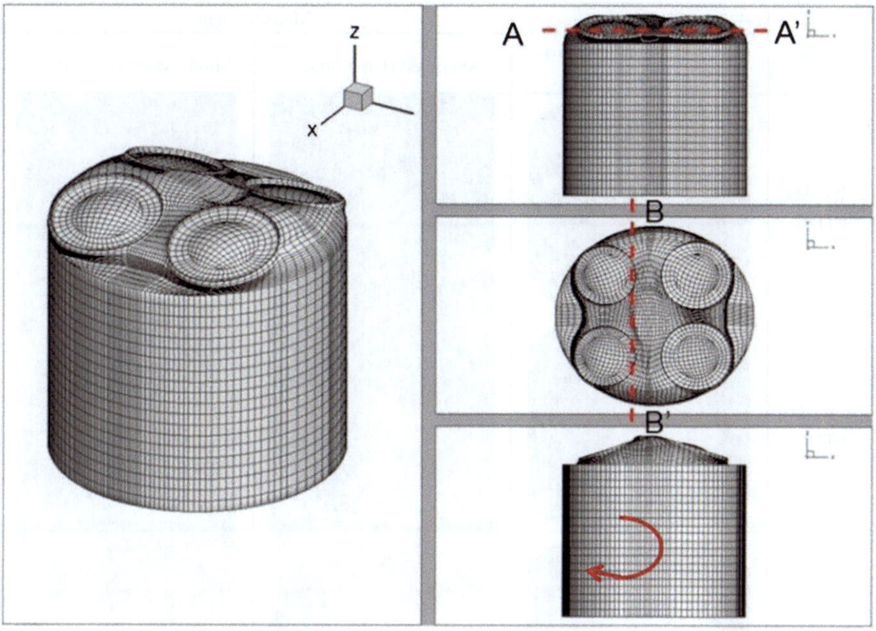

Fig. 1 Computational mesh with 62 756 cells for the research engine at IVC. Section A-A′ horizontally cuts through a point located at 2 mm above the top of cylinder linear. Section B-B′ vertically cuts through the center of the hollow-cone spray

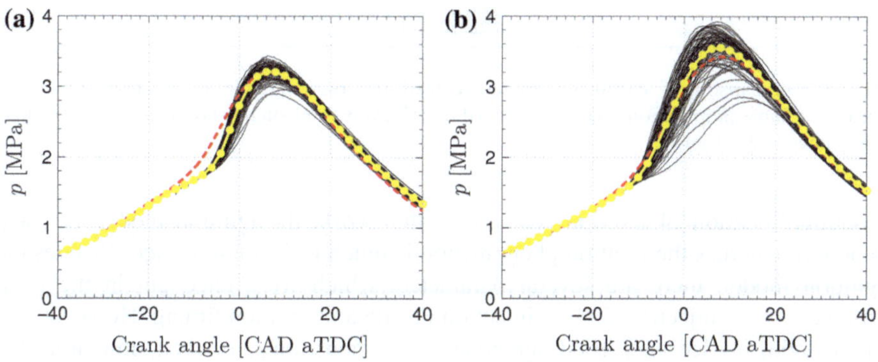

Fig. 2 Comparison between the calculations (*red dashed lines*) with measurements (cycles 1–100: *black solid lines*; ensemble-averaged: *yellow solid line* with *circles*). **a** low load: $t_{inj} = -20$ CAD aTDC, $t_{ign} = -16$ CAD aTDC, $\lambda = 2.6$, imep = 3.8 bar; **b** medium load: $t_{inj} = -32$ CAD aTDC, $t_{ign} = -20$ CAD aTDC, $\lambda = 1.85$, imep = 5 bar

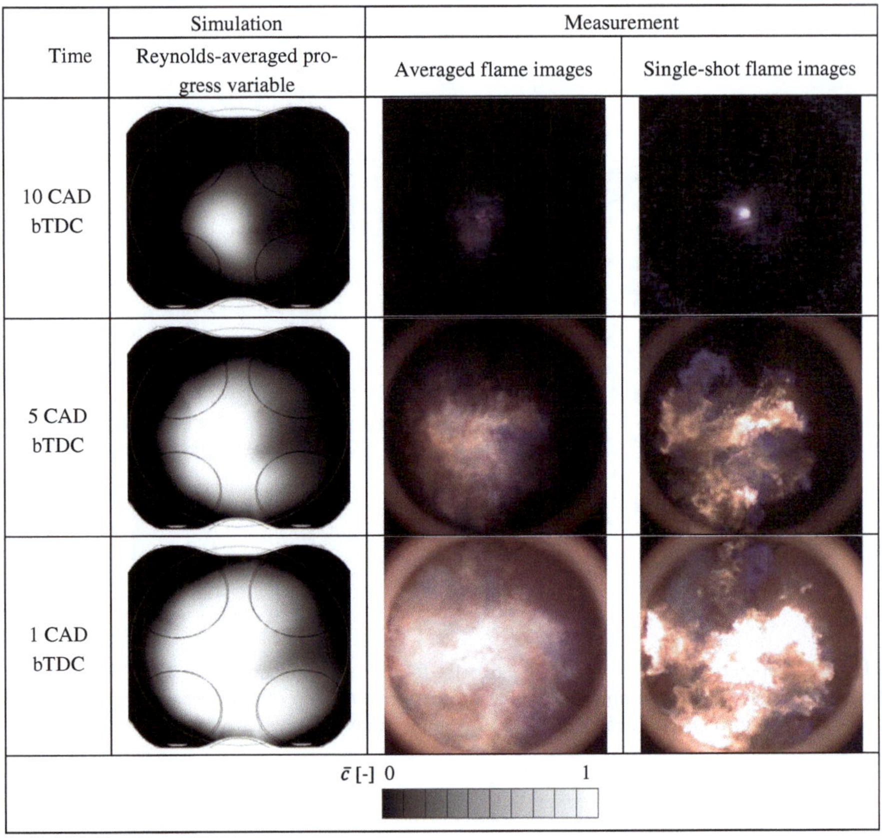

Fig. 3 Comparison of calculated Favre-averaged progress variable (in section A-A′) and measured flame propagation images (averaged over 7 cycles and single-shot images) for low load case (imep = 3.8 bar)

oversimplification of ignition process. In this work, the ignition model is a step function, whereas the ignition phenomenon is much more complicated. Besides the ignition model, there are several parameters which were uncertain in the measurements, e.g. injection timing, injection profile and ignition timing. However, it's encouraging to see very good agreement in combustion rate when simulating the experiments at different loads with very short time intervals between injection and ignition.

Comparison of the computed Reynolds-averaged progress variable $\bar{c}$ at section A-A′ and measured averaged flame propagation images over 7 cycles as well as single shot images is shown in Fig. 3. It is worth clarifying the physical meaning of Reynolds-averaged progress variable; it corresponds to the probability of finding the burned products. At the beginning of the burning stage, i.e. 10 CAD bTDC, a larger flame kernel at the spark position is predicted by the simulation as compared to that in the measurement. Such difference in flame kernel size might be caused by the use of a simple ignition model in the simulation. Later on, the turbulent flame continues to expand to the peripheral of the combustion chamber with a preferred direction to the left of the combustion chamber due to the location of the spark plug. At 1 CAD bTDC, a complete combustion is observed at the lower right part of computed image whereas very weak combustion is captured in measured one. This discrepancy might be caused by the formation of soot on the glass piston which blocks the optical view in some of the measured images. In addition, single-shot flame images are presented in order to show a more clear flame structure as compared to the averaged flame images. It is worth stressing that the experimentally observed images contain not only information of flame location but also spectroscopic information of soot location. It can be seen that the URANS simulation resembles the single-shot flame images very well.

5 Conclusions

The fuel injection by an outward-opening piezo-actuated pintle-type injector and stratified turbulent combustion in a SG GDI engine was modeled by using an open source code. Good agreement was obtained by comparing the pressure traces between the simulations and measurements at two load points especially for the pressure rise rate. Both calculated Reynolds-averaged progress variable and averaged flame propagation images showed that the burning was mainly located in the right side of the combustion chamber and this was due to the location of the spark.

Acknowledgments This work was supported by the Combustion Engine Research Center (CERC) and the Swedish Energy Agency. We are very grateful to Dr. A. Imren for providing the computational mesh and KIVA simulation, to Dr. F. Kärrholm, Dr. T. Lucchini, and Dr. N. Nordin for kindly providing training in the use of the OpenFOAM library, to Prof. I. Denbratt, Dr. V. Golovitchev, and Dr. A. Karlsson for valuable discussions, and to C^3SE at Chalmers for providing computational resources.

References

1. European Vehicle Market Statistics (2013). http://www.theicct.org/sites/default/files/publications/EU_vehiclemarket_pocketbook_2013_Web.pdf. Accessed 6 Aug 2014
2. EPA Staff Technical Report (2008) United States Environmental Protection Agency. http://www.epa.gov/otaq/climate/420r08008.pdf. Accessed 8 Aug 2014
3. Wirth et al (2003) The next generation of gasoline direct injection: improved fuel economy and optimized system cost, 12. AACHENER KOLLOQUIUM FAHRZEUG- UND MOTORENTECHNIK, 6.–8. OKTOBER 2003
4. Drake MC, Haworth DD (2007) Advanced gasoline engine development using optical diagnostics and numerical modeling. Proc Combust Inst 31(1):99–124. doi:10.1016/j.proci.2006.08.120
5. Huang C, Lipatnikov A (2014) A numerical study on stratified turbulent combustion in a direct-injection spark-ignition gasoline engine using an open-source code. SAE 2014-01-1126

6. Huang C, Lipatnikov A (2014) A study of two basic issues relevant to RANS simulations of stratified turbulent combustion in a direct injection spark ignition engine. SAE 2014-01-2572
7. Huang C, Lipatnikov A (2011) Modelling of gasoline and ethanol hollow-cone sprays using OpenFOAM. SAE paper 2011-01-1896
8. Hemdal S, Denbratt I, Dahlander P, Warnberg J (2009) Stratified cold start sprays of gasoline-ethanol blends. SAE Int J Fuels Lubr 2(1):683–696
9. Ogink R, Golovitchev V (2002) Gasoline HCCI modeling: an engine cycle simulation code with a multi-zone combustion model. SAE paper 2002-01-1745
10. Huang C, Golovitchev V Lipatnikov A (2010) Chemical model of gasoline-ethanol blends for internal combustion engine applications. SAE paper 2010-01-0543

Ignition Delay, Lift-off and Soot Luminescence in Diesel-Ethanol Spray Combustion

Chengjun Du, Mats Andersson and Sven Andersson

Abstract To investigate the influence of ethanol blending in Diesel fuel on the spray combustion characteristics, experiments were carried out in an optically accessed high-pressure/high-temperature spray chamber. Three different fuels were investigated: (1) Diesel, based on the European EN590 standard; (2) E10 with 10 % ethanol and (3) E20 with 20 % ethanol. Gas conditions were kept at a constant gas density of 24.3 kg/m^3 at a lower combusting temperature (550 °C, 57.3 bar) and a higher combusting temperature (600 °C, 60 bar). The experimental results showed that ignition delay and delay time of soot formation were increased as ethanol was blended in Diesel fuel. Less soot luminescence was observed and longer lift-off length was measured in Diesel-ethanol blends compared to pure Diesel.

1 Introduction

Mixing a renewable fuel into conventional fossil fuels enables a reduction in net green-houses gas emissions, and blending of ethanol into Diesel is one such alternative. Furthermore, it has been reported that a reduction of particle emissions can be achieved using Diesel-ethanol blends in a Diesel engine [1–3]. However, little work has been done in a spray chamber to investigate the combusting spray characteristics of Diesel-ethanol blends, which could give more insights into the spray development, ignition and combustion of Diesel-ethanol blends compared to Diesel. Therefore, the influence of fuel composition on combustion characteristics was studied, including ignition delay, and the time evolution of OH^* chemiluminescence, soot luminescence and lift-off length.

C. Du (✉) · M. Andersson · S. Andersson
Department of Applied Mechanics, Chalmers University of Technology, Gothenburg, Sweden
e-mail: Chengjun.Du@chalmers.se

I. Denbratt et al. (eds.), *Sustainable Automotive Technologies 2014*,
Lecture Notes in Mobility, DOI 10.1007/978-3-319-17999-5_8

2 Experimental Apparatus

Three different fuels were used, Diesel based on the European EN590 standard, E10 (10 % Ethanol and 88 % Diesel) and E20 (20 % Ethanol and 78 % Diesel). A 2 % additive was used to create appropriate solubility and lubricity for each Diesel-ethanol blend. A Bosch solenoid injector with a single-hole nozzle (diameter 0.14 mm) was used and the injection pressure was kept at 1350 bar. The injection duration was 3.5 ms. The gas density in the spray chamber was kept constant at 24.3 kg/m^3 for all of the experiments, corresponding to 550 °C at 57.3 bar and 600 °C at 60 bar. Several optical methods were used to measure various spray characteristics as shown in Fig. 1. To study the spray process, shadow images with a high speed video camera were captured. OH^* chemiluminescence images and soot luminescence images were acquired with intensified CCD cameras using a 308 nm bandpass filter with a 10 nm FWHM and 630 nm long pass filter between the respective camera and a dichroic beam splitter.

The start of high-temperature chemistry (i.e. ignition delay) and the start of soot formation (i.e. delay time of soot formation) were detected by photomultiplier tubes

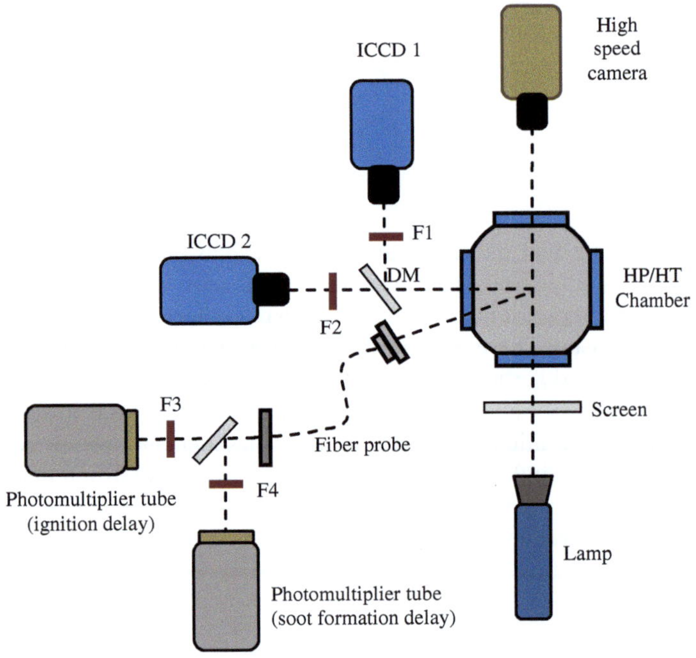

Fig. 1 Schematic of the experimental setup for the shadowgraph imaging, capturing images of OH-chemiluminescence and soot luminescence, detecting delay time of soot formation and ignition delay. *DM* dichroic mirrors, *F1* 308 nm (10 nm FWHM), *F2* RG630, *F3* 300 nm (25 nm FWHM), *F4* RG610

(PMT) collecting light from the combusting spray. The PMTs, with optical filters selecting OH^* and soot luminescence light, detected the light intensity as a function of time.

3 Results and Discussion

3.1 Spray Development

Figure 2 shows the whole spray development using Diesel, E10 and E20. At the second image in each series, which corresponds to 0.3 ms after start of injection (ASOI), maximum liquid phase penetration has already been established and keeps relatively stable, and there is no big difference on the liquid spray penetration between Diesel, E10 and E20 [4]. After some time a bright flame caused by soot incandescence appears and it can clearly be seen that it occurs first for Diesel and last for E20.

3.2 Delay Time of Ignition and Soot Formation

For measuring the delay time of ignition and soot formation, over 300 individual injection events were performed for each fuel, and an averaged UV luminosity and

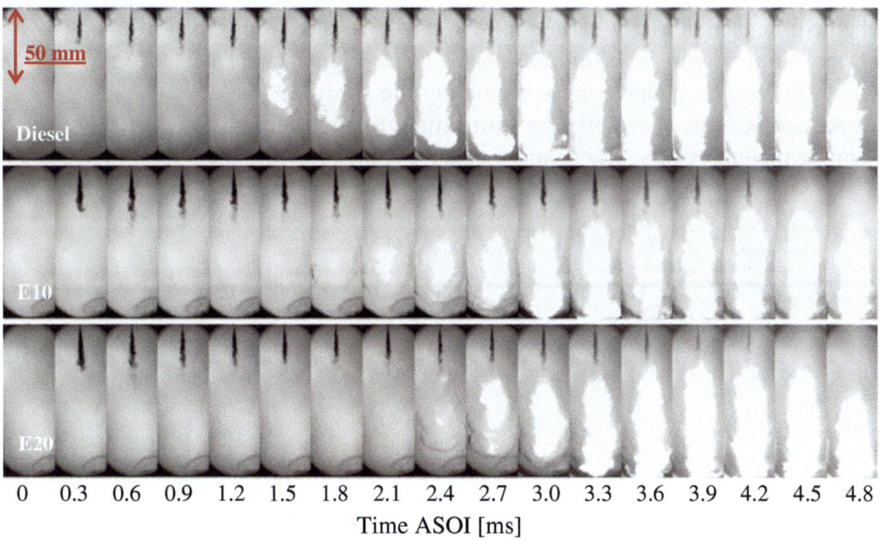

Fig. 2 Selected images from video sequences showing the spray development using Diesel (*top* row), E10 (*middle*) and E20 (*bottom*). Experimental conditions: ambient temperature 600 °C, ambient pressure 60 bar

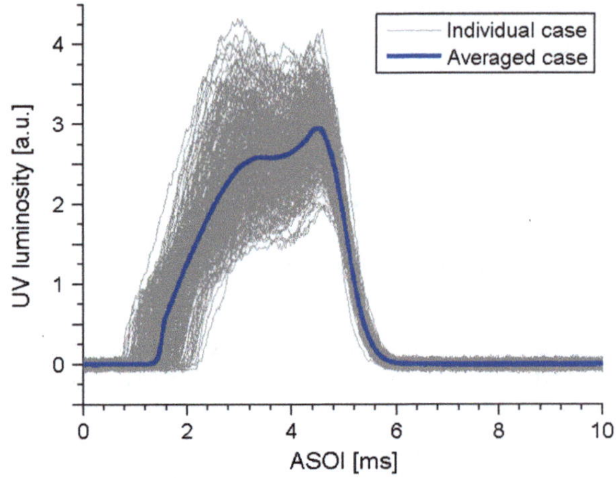

Fig. 3 UV luminosity measurements for Diesel under over 300 individual injection events at ambient temperature 600 °C, ambient pressure 60 bar. Thick blue line is a conditioned-average as described in the text

soot luminosity were obtained. Figure 3 shows individual UV luminosity curves and the average UV luminosity (thick blue line). The latter is a combination of a conditioned-average in the initial part of the curve, till it reaches an intensity of 0.5 a.u., and regular average after that. The conditioned-average was obtained by time shifting the curves so that they reached an intensity value of 0.5 a.u. at the mean time required to reach this value, the same method was used as in [5]. By doing so, the delay time of ignition could be evaluated more accurately.

Start timing of UV luminosity (i.e. ignition delay) is later as more ethanol is blended in Diesel fuel as shown in Fig. 4a, which means that high-temperature chemistry reaction starts earlier for Diesel. It results from ethanol having lower cetane number, and in addition, ethanol has a higher latent heat of vaporization, which lowers the temperature surrounding the spray.

As can be seen in Fig. 4b, the start timing of soot luminosity (i.e. delay time of soot formation) occurs later as more ethanol is blended into the Diesel fuel. Start of

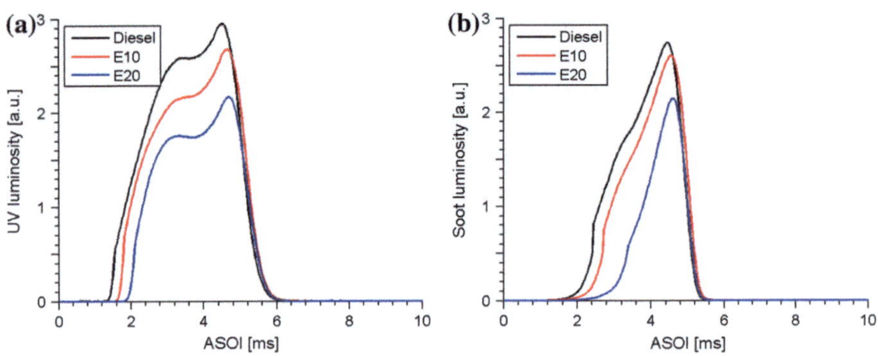

Fig. 4 Average UV (**a**) and soot (**b**) luminosity for Diesel, E10 and E20. Experimental conditions are ambient temperature 600 °C, ambient pressure 60 bar

Table 1 Start time ASOI of UV and soot luminosity at different combusting temperature

	Ambient temperature 550 °C		Ambient temperature 600 °C	
	UV luminosity (ms)	Soot luminosity (ms)	UV luminosity (ms)	Soot luminosity (ms)
Diesel	1.78	2.10	1.46	2.06
E10	2.84	3.08	1.72	2.34
E20	3.08	3.24	2.00	2.90

soot formation is later than ignition for all three fuels, with the longest delay for E20 among them.

In Table 1 it can be seen that when the ambient temperature is lowered from 600 °C to 550 °C, the ignition delay becomes longer, and the difference between the Diesel fuel and the Diesel-ethanol blends become larger. Lowering the temperature has a smaller effect on the delay time for soot formation, except for the E10 fuel.

3.3 Lift-off Length

Lift-off lengths are evaluated based on OH^* chemiluminescence images taken at various times ASOI, and the methodology of evaluating lift-off length is explained in [4]. The results shown in Fig. 5 are based on averaged images from 30 injection events at each time step. At 550 °C, the order for reaching relatively stable lift-off length is Diesel, E10 and E20, and also the order of short to long lift-off length is Diesel, E10 and E20, which is in agreement with fuels with a lower cetane number having longer ignition delays and longer lift-off lengths [6]. It is also evident for E10 and E20 that the lift-off length develops from downstream to upstream of the spray and then stabilizes at the shortest lift-off length.

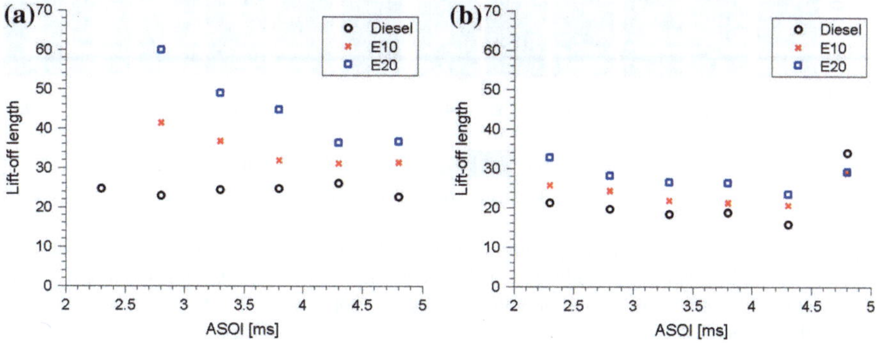

Fig. 5 Lift-off length development with time. Experimental conditions: **a** Ambient temperature 550 °C and pressure 57.3 bar, **b** Ambient temperature 600 °C and pressure 60 bar

3.4 Soot Luminescence

Soot luminescence images were taken simultaneously with OH^* chemiluminescence images, and the images in Fig. 6 are averages of 30 images taken at the each time ASOI. The ICCD camera gain and gate time were kept identical for all the fuels. As shown in Fig. 6a, Diesel has the earliest soot luminescence, which agrees well with the start of soot formation measured with the PMT, shown in Table 1. Regardless of ambient temperature, as ethanol contents increases in the fuel, there is less soot luminescence intenstiy at the same time ASOI, which indicates that less soot is formed in the flame of Diesel-ethanol blends. The decrease in soot with increasing ethanol contents correlates well with an increase in ignition delay and longer lift-off length as described above. As the lift-off length becomes longer, more air can be entrained towards the center of the spray, which leads to less rich conditions in the flame and less soot is formed. Furthermore, oxygenated fuels have been shown to produce less soot [7].

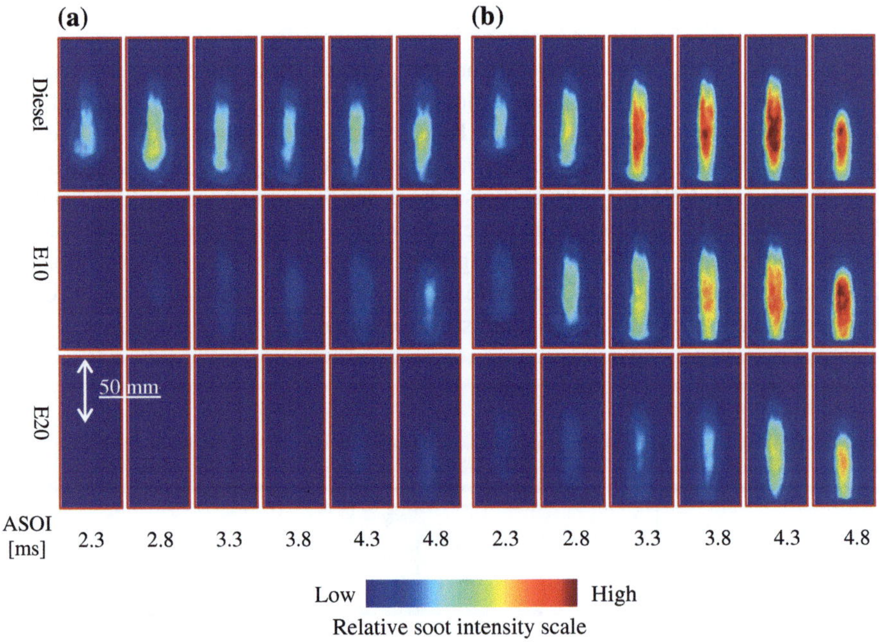

Fig. 6 Soot luminescence development with time. All the images are in the same colormap intensity scale. Experimental conditions: **a** Ambient temperature 550 °C and pressure 57.3 bar, **b** Ambient temperature 600 °C and pressure 60 bar. The upper edge of the images corresponds to the nozzle tip location

4 Conclusions

Spray experiments were carried out with Diesel-ethanol blends to investigate spray characteristics under combusting conditions. The following conclusions can be reached in this study as the ethanol content is increased from 0 % to 20 %:

- Ignition delay and delay time of soot formation increases, and there is delay from ignition to start of soot formation for all fuels;
- The lift-off length becomes longer and it takes a longer time to establish a stable lift-off length;
- Less soot luminescence is detected at the same time ASOI.

Acknowledgments Financial support was obtained from the Combustion Engine Research Centre (CERC) at Chalmers. The fuel blends were provided by Dr Anna Lif at AkzoNobel.

References

1. Hu C, Jian-Xin W, Shi-Jin S, Xin-Liang A, Wen-Miao C (2006) SAE Technical paper 2006-01-0236
2. Cui X, Helmantel A, Golovichev V, Denbratt I (2009) SAE Technical paper 2009-01-2695
3. Rakopoulos DC, Rakopoulos CD, Kakaras EC, Giakoumis EG (2008) Energy conversion and management, vol 49, pp 3155–3162
4. Du C, Andersson M, Andersson S (2014) The influence of ethanol blending in diesel fuel on the spray and spray combustion characteristics. SAE Int J Fuels Lubr 7(3):823–832, doi:10.4271/2014-01-2755
5. Higgins B, Siebers D, Aradi A (2000) SAE Technical Paper 2000-01-0940
6. Pickett LM, Siebers DL, Idicheria CA (2005) SAE Technical Paper 2005-01-3843
7. Ito T, Kitamura T, Ueda M, Matsumoto T, Senda J, Fujimoto H (2003) SAE Technical paper 2003-01-0073

Lean Upgrade of Aftertreatment Systems to Euro6b Compliance

P. Nilsson, C. Wang-Hansen, M. Lundgren and M. Hicks

Abstract Meeting new emission standards can be a challenging task, where many optimization parameters often lead to a variety of hardware designs and engine control strategies. Moreover, the life cycle of an exhaust gas aftertreatment system (EATS) is much shorter compared to the engine. Resource demanding and costly engineering programs are therefore common. The aftertreatment upgrade to Euro6b compliance for the 2.4 l five-cylinder Volvo Environmental Diesel engine (VED5) represents an exception. Based on two fundamentally different metalwork designs for front wheel- and all-wheel drive (FWD and AWD respectively) vehicles and carefully adapted control strategies, the Euro5 base design was transferred with minimal hardware changes.

1 Introduction

The efficiency and complexity of exhaust aftertreatment systems (EATS) for diesel engines have increased significantly over the past 15 years resulting in a considerable increase in development- and hardware costs, as well as tooling investments. Changes have been primarily driven by stepwise introduction of new emission legislations. However, the EATS design has also been influenced by continuous engine development till today's 2.0 l four-cylinder engine family (DRIVe powertrains), characterized by high specific performance and low fuel consumption [1]. Volvo Car Corporation (VCC) has been in the frontline regarding EATS technology by not only adapting efficiently to emission limits, but also quickly introducing the required technology to meet engine engineering targets [2, 3].

The first in-house developed catalytic converter, the Diesel Oxidation Catalyst (DOC) was introduced together with the all new VCC market leading 2.4 l in-line five-cylinder diesel engine in 2001. Four years later VCC introduced the Diesel

P. Nilsson (✉) · C. Wang-Hansen · M. Lundgren · M. Hicks
Volvo Car Corporation, Gothenburg, Sweden
e-mail: pnilss33@volvocars.com

I. Denbratt et al. (eds.), *Sustainable Automotive Technologies 2014*,
Lecture Notes in Mobility, DOI 10.1007/978-3-319-17999-5_9

Particulate Filter (DPF). The consecutive emission limit (Euro5 in 2009) was met by significant controls and calibration improvements, e.g. filter regeneration functionality and strategy [4]. During the same period of time, the engine was significantly reworked. The new class leading engine, distinguished by a very attractive balance between performance and fuel economy, was launched in 2008. The twin-turbo charging system resulted in changed prerequisites for DPF regeneration (by a reduction in exhaust gas temperature and increase in mass flow), which contributed to the introduction of a close-coupled (CC) DPF installation in 2010.

1.1 Lean NOx Trap Technology

The current upgrade of the VED5 exhaust aftertreatment system from Euro5 to 6b requires more than 50 % NOx reduction over the New European Drive Cycle (NEDC). Technologies with the potential to meet these targets include Lean NOx Traps (LNT) and urea based Selective Catalytic Reduction (SCR). LNT is clearly the technology of choice, as the complex and expensive on-board urea infrastructure required for the SCR system is not needed. Moreover the LNT provides much greater low temperature efficiency and lower system backpressure. The decision to develop the LNT technology was taken in 2010, with the first introduction in the all-new four-cylinder 2.0-l engine family—the DRIVe powertrains—in autumn 2013.

NOx reduction by the LNT is based on several kilometers of lean operation, during which NOx is adsorbed in the storage compound, followed by a few seconds of rich operation, where stored NOx is released and reduced by mainly CO and H_2 (deNOx). The necessary quantity of reductants, i.e. the sum of adsorbed NOx and oxygen storage capacity (OSC) with consideration to the catalyst efficiency, is given by the rich duration and lambda target. Specific engine control strategies required to generate rich conditions were first developed for the DRIVe powertrains which was used as a base for the current VED5 upgrade. Another general challenge, where we could rely on knowledge gained in the first LNT project, was to understand and delimit sulfur deactivation. Adsorption of SO_2 (originating from diesel fuel and oil additives) leads to a loss of NOx storage capacity. This sulfur poisoning is largely reversible, i.e. the adsorption capacity can be restored by sulfur regeneration (deSOx), but a minor progressive degradation needs to be taken into account by dedicated engine control strategies.

1.2 Objectives

To meet efficiency targets, requires not only efficient NOx and sulfur regeneration, but an appropriate amount of catalyst material placed in a suitable temperature window. The FWD metalworks design limits the available LNT volume, and the application for which it is intended—the Plug-in Hybrid Electrical Vehicle (PHEV) —produces a challenging temperature and emissions profile. On the contrary, flexibility regarding LNT volume is much greater for the all-wheel drive (AWD) application, where the main challenge is instead to limit high peak temperatures in the LNT during DPF regeneration.

2 System Design History

This chapter is intended to give the reader an understanding of the critical design parameters of the current Euro5 systems for FWD and AWD applications which are the base for the upgrade to Euro6b. The metalworks design can be described by four steps; (1) estimation of required catalyst volume, (2) assessment of installation limitations, (3) maximization of Uniformity Index (UI) and minimization of backpressure (e.g. by adjusting substrate contour, cone angles, pipe diameters, etc.) and finally, (4) determination of optimal sensor placement.

Based on Euro5 emission limits and engine-out emissions, a one liter DOC followed by a 3 l DPF was optimal for FWD vehicles. The DPF was placed in CC position to minimize heat losses and thereby improve passive and active regeneration efficiency (see Fig. 1a). In AWD vehicles the Power Transmission Unit (PTU) and prop shaft makes a close-coupled DPF installation impossible, as shown in Fig. 1b, where the prop shaft is colored in red. A 1.5 l DOC is used since HC and CO emissions are higher compared to FWD vehicles and since a DPF in underfloor position increases thermal aging. The AWD DPF, which is a carry-over design (i.e. contour wise) from Euro4 EATS, is 4.3 l.

After consideration to packaging restraints, Computational Fluid Dynamic (CFD) simulations were performed in several design loops to optimize the design. This was done to obtain an even flow distribution of the exhaust gas, i.e. a Uniformity Index (UI) close to one, and a low system backpressure. High UI is important to obtain full utilization of the catalysts and limit catalyst aging. An example of a velocity profile for the AWD system, is shown in Fig. 2a. Backpressure of an aftertreatment system is an important property. Too high backpressure is negative for max power and fuel efficiency of the engine.

CFD simulations were also conducted to find optimal positions for the Exhaust Gas Temperature Sensors (EGTS). An example for the positioning of rear temperature sensor in AWD system is shown in Fig. 2b. This resulted in improved temperature control during DPF regenerations, and improved accuracy for the DPF soot load estimation compared to previous designs. As a result fuel penalty connected to active DPF regeneration was minimized and engine and aftertreatment durability was increased.

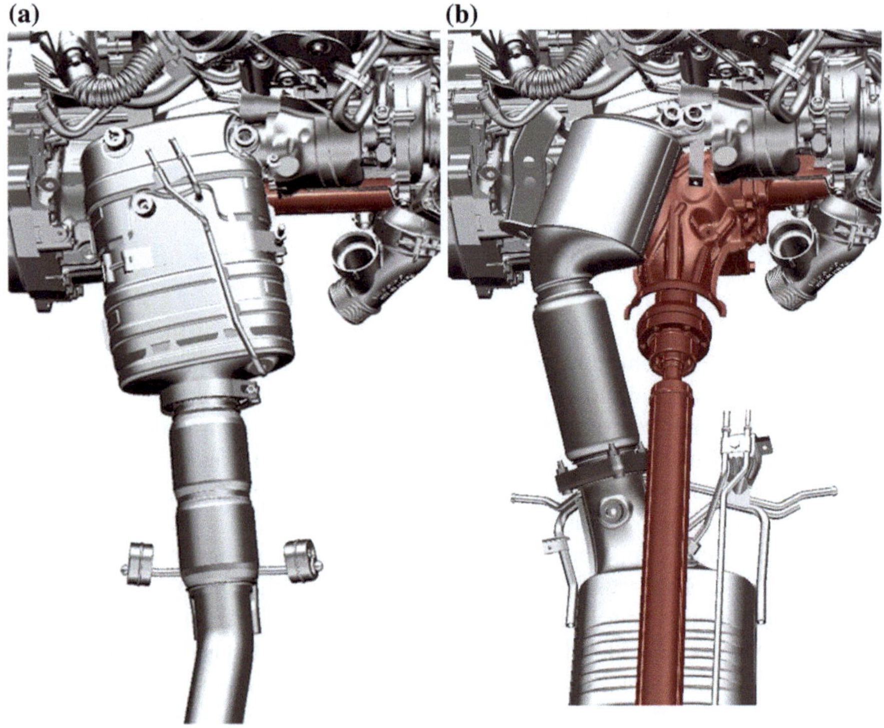

Fig. 1 Packaging prerequisites for FWD (**a**) and AWD (**b**) applications. PTU and/or prop shaft is highlighted in *red*. The FWD CC unit contains DOC and DPF in an in-line arrangement. The AWD DOC is placed in front of the PTU, and the inlet half of the DPF is shown at the *bottom*

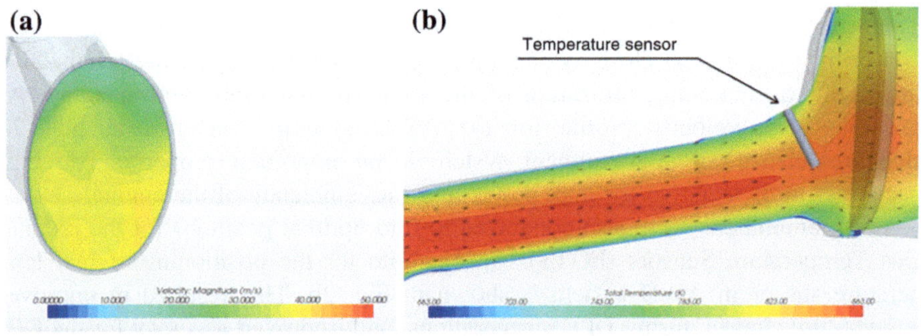

Fig. 2 Velocity profile for DOC in AWD EATS from CFD simulations (**a**). Example of CFD heat flow simulation to position rear temperature sensor for AWD system (**b**)

3 Euro6b Development and Solutions

The legal upgrade from Euro5 to Euro6b does not include changes to the VED5 base engine. This significantly simplifies the development of new aftertreatment systems. At the start of the project, engine-out emissions were at reasonable levels and the engine reliability was good. Moreover, the target was to keep the Euro5 systems design as far as possible and thereby profit from the development work done previously, described in Sect. 2. The only change in the metalworks design, was the addition of a second lambda sensor (and a new Engine Control Unit) to control rich-mode operation during deNOx and deSOx. In an early project phase, demonstrator vehicles were built. These were used to show the aftertreatment systems potential by measuring catalyst temperatures and conversion efficiencies.

3.1 PHEV System

The Euro6b application of the Euro5 FWD design is a Plug-in Hybrid Electrical Vehicle (V60 car). The PHEV application means new operating patterns for the diesel engine. In hybrid mode, the electric motor (rear axle) is used during low load driving and the diesel engine cuts in at higher loads. In the NEDC, this means that the diesel engine is started and stopped several times and battery charge neutrality is kept over the cycle. As a result, high transients in both engine-out temperature and NOx emissions occur, and heat-up time for the engine is longer. On the other hand, it is not used for engine braking, which means there is no cooling effect from pumping air through the EATS in over-run.

The washcoat load is higher for the LNT (compared to the previous DOC) to obtain the required NOx storage capacity. Since the thickness of the washcoat is limited by the internal diffusion rate, the substrate volume must be larger. The DOC in the Euro5 system is 1 l, and a 1.5 l LNT can be considered baseline to fulfill Euro6b NOx targets. However, as the DPF optimally should fit in the same unit, the available volume was redistributed to 1.5 l LNT and 2.5 l DPF (indicated in Fig. 3).

But as the velocity profile entering the LNT is somewhat uneven, the effective volume is less than 1.5 l. With the first LNT project as a reference, this is compensated for by an increase in PGM (Platinum Group Metals) load and by the introduction of next generation LNT washcoat technology. Main advancements are greater NOx storage capacity and lower desulfation temperature, as shown in Fig. 4. In addition, a lower desulfation temperature contributes to less thermal ageing of the LNT.

The reduction in DPF volume can have a negative impact on ash storage capacity, which is partly compensated by a reduction in wall thickness (from 12 to 10 mil). However, since the diesel engine operating time during the vehicle life is shorter in the PHEV application, less ash storage capacity is needed. Another change to the DPF was the introduction of a specific H_2S clean-up washcoat, which

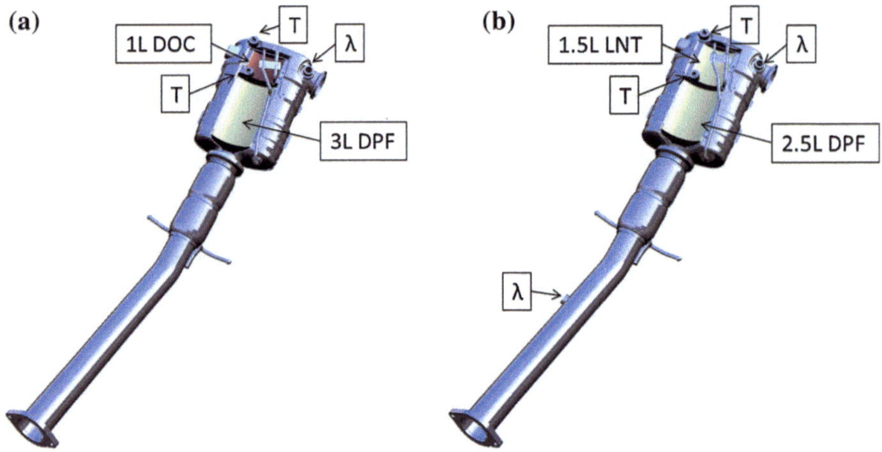

Fig. 3 Overview of Euro5 FWD design (**a**) and Euro6b PHEV design (**b**) showing the rear lambda sensor position and redistributed front-/rear brick volume

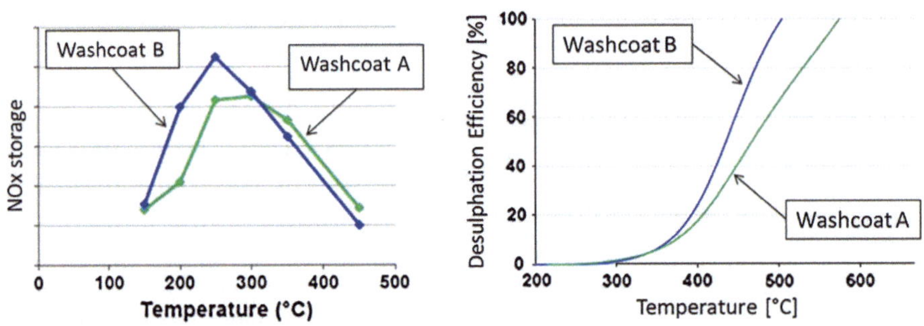

Fig. 4 Improvements in NOx storage capacity (*left*) and desulfation efficiency (*right*) for 1st generation (A) and 2nd generation (B) LNT washcoats

takes care of hydrogen sulfide that may form over the LNT during sulfur regeneration.

In summary changes to the first substrate include: increased length and new washcoat (with significantly higher washcoat load) and changes to the DPF were: reduced length, reduced wall thickness and new washcoat formulation. If the complete system design is not carefully balanced, backpressure can easily rise for an LNT based Euro6b system, compared to a Euro5 system. Here the higher backpressure of the first substrate was effectively compensated for, by the changes made to the DPF (Fig. 5).

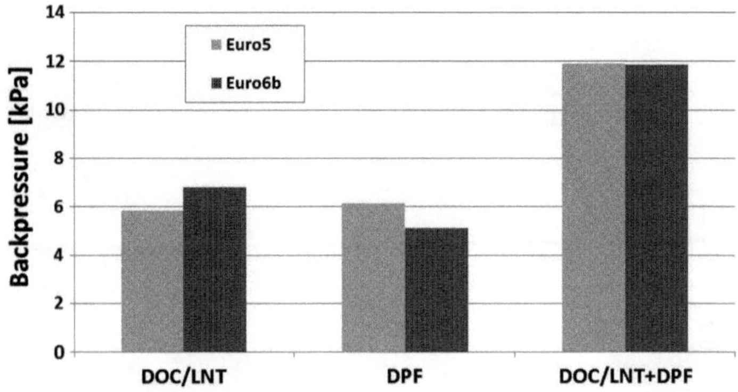

Fig. 5 Coldflow (800 kg/h, 20 °C) backpressure of the PHEV system

3.2 AWD System

The Euro6b AWD system is intended for vehicles within D-segment (e.g. XC60). These applications are characterized by higher exhaust flow and -temperature and higher emission levels, compared to corresponding FWD vehicles. The exhaust aftertreatment system must have higher conversion efficiency and greater resistance to thermal aging as a result of the underfloor location of the DPF.

The CC unit was not changed as the available substrate volume of 1.5 l was considered to be sufficient. A prerequisite was the optimized inlet cone design, resulting in an even flow distribution and low thermal loss from turbo to LNT. The ceramic 400/4 substrate was also kept, since it supports the LNT functionality well. The rear lambda sensor, needed for deNOx and deSOx control, was placed in the inlet cone of the DPF instead of directly after the LNT due to installation limitations (see Fig. 6b). Sensor position is generally a balance between response time and gas mixing distance to provide quick and accurate response. The current placement somewhat favors accuracy.

To ensure efficient conversion at aged conditions, 2nd generation LNT washcoat technology and the higher PGM load (similar to the PHEV system) were used also for this application. Figure 7 shows the superior NOx conversion efficiency over the NEDC for the 2nd generation technology compared to 1st generation. Also note the high efficiency in the low speed/low temperature part of the cycle which is characteristic to the LNT technology. Moreover, active DPF regeneration strategies were optimized to reduce thermal load on LNT. Temperature overshoots were reduced by improved temperature control, and active regeneration target temperatures, for both low and high soot loads, were rebalanced.

The UF unit was originally designed for a 12″ substrate (Euro4). The system upgrade to Euro5 included a first length reduction to 10″, which was made possible by advancement in active regeneration control. The current Euro6b system upgrade includes a second length reduction to 8″, enabled by advancements in substrate

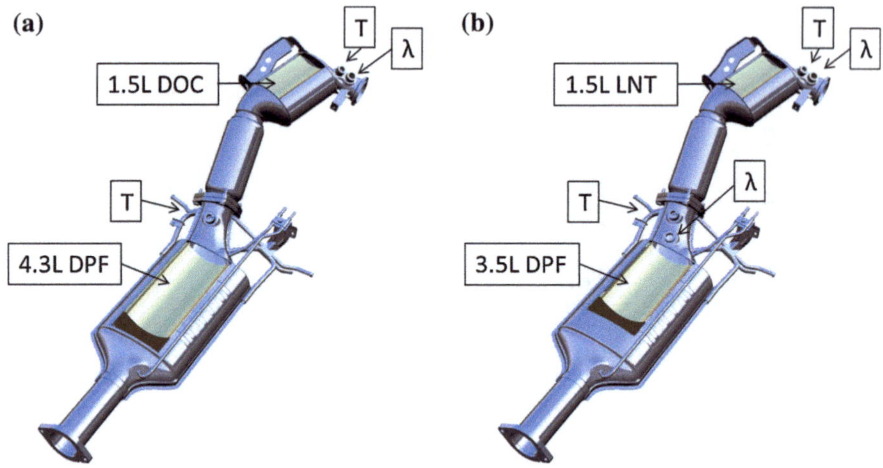

Fig. 6 Overview of AWD Euro5 design (**a**) and Euro6b design (**b**) showing the rear lambda sensor position and shortened DPF

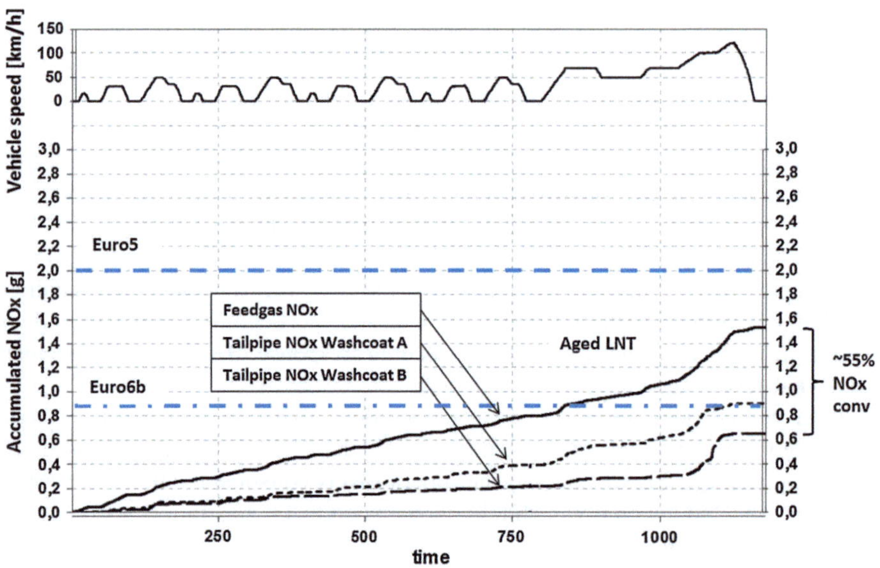

Fig. 7 Accumulated engine-out (feedgas) NOx and post-LNT NOx for first (A) and second (B) generation LNT washcoat over the NEDC cycle for the AWD system

technology (Fig. 6). The challenge was to make the change without influence on emission performance or backpressure. Substrate changes included: a transition from symmetric- to asymmetric channels, a reduction in wall thickness (12 to 10 mil) and an increase in porosity. Naturally this also imposed changes to the DPF washcoat: the load was reduced, the washcoat and PGM zoning was changed, and the active ingredient for H_2S control was changed. Final backpressure and emission performance were well within targets.

4 Concluding Remarks

The legal upgrade to Euro6b for the VED5 exhaust aftertreatment system was performed with intact metalwork design by changing catalytic internals and a reusing knowledge and experience gathered in previous and parallel projects. The PHEV system is based on the Euro5 front-wheel drive installation. The main changes include: redistribution of first and second brick volume in the close-coupled unit, application of 2nd generation LNT washcoat technology and reduction of DPF substrate wall thickness and modified DPF washcoat.

The Euro5 all-wheel drive installation is of conventional design with a close-coupled DOC followed by a DPF in underfloor position. This system layout offers a high degree of flexibility regarding catalyst volume, but requires sophisticated regeneration strategies including soot estimation models and temperature control. For Euro6b, the DOC was replaced by an LNT, with the same substrate specification. DPF volume could be reduced, by using a substrate with asymmetric cell structure, without compromising service life demands. In summary, with the above mentioned changes, a significant reduction in NO_x tailpipe emissions was accomplished for both PHEV and AWD vehicles.

References

1. Crabb DC, Fleiss M, Larsson J-E, Somhorst JHE (2013) VEA—Volvo Engine Architecture—the Volvo answer to a challenging market: extreme downsizing and maximum commonality while maintaining highly competitive customer attributes. MTZ worldwide Edition: 2013-09
2. Laurell M, Sjörs J, Ovesson S, Lundgren M, Brück R, Presti M (2013) The Innovative exhaust gas aftertreatment system for the new Volvo 4 cylinder engines: a unit catalyst system for gasoline and diesel cars. In: Proceedings of 22nd Aachen colloquium automobile and engine technology
3. Johansson C, Laurell M, Brück R (2009) Exhaust gas treatment device for use near engine. Pat. DE 10 2009 024 718, Germany
4. Lundgren M (2009) Light duty diesel exhaust aftertreatment from an OEM point of view. In: Proceedings of 7th international CTI forum exhaust systems. Düsseldorf, Germany

Delivery Evaluation of High Pressure Natural Gas Fuel Injection

Thomas Rogers, Phred Petersen, Prashanth Karra and Petros Lappas

Abstract Schlieren high speed videography was employed to ascertain the geometric properties, compressible flow behaviour and the flow field of Compressed Natural Gas (CNG) jets. Axial and radial penetration, velocity, and dispersion have been studied comparatively. Schlieren high speed imaging reveals the dynamic flow behaviour of the delivery event, namely barrel Mach disc formation and mixing at the shear layer alongside other geometrical features; spreading rate, self-similarity, velocity and penetration. New findings of high-pressure gaseous fuel delivery are presented based on observations in a constant-volume chamber that replicates engine conditions. A novel analysis method for characterising structural dynamic properties of the gas jet are also presented alongside a new delivery method.

1 Introduction

The use of compressed natural gas (CNG) is on the rise due to its relative abundance, lower combustion emissions and economic viability. Gaseous fuel injector development for direct injection spark ignition (DISI) engines is still largely in its infancy. Whilst several commercial options have appeared in recent years, there is still a relatively large amount of knowledge required on the delivery and mixing processes particularly in the nearfield at low after start of injection (aSOI) times.

The aim of this work is to highlight the structural properties of free-stream and impinging compressed natural gas jets delivered into a quiescent direct-fuel injection environment. High pressure, large pressure-ratio, fuel delivery is sought to decrease delivery time and increase the energy delivered for direct-injection

T. Rogers (✉) · P. Karra · P. Lappas
School of Aerospace, Mechanical and Manufacturing Engineering, RMIT University, Melbourne, Australia
e-mail: Thomas.Rogers@rmit.edu.au

P. Petersen
School of Media and Communication, RMIT University, Melbourne, Australia

I. Denbratt et al. (eds.), *Sustainable Automotive Technologies 2014*,
Lecture Notes in Mobility, DOI 10.1007/978-3-319-17999-5_10

spark-ignition engines. The intent of this work is to capture the delivery process using both qualitative and quantitative methodologies, highlighting the various dynamic phenomenological aspects of these flow conditions. One key objective in this work is to ascertain the transient development in the starting region of the CNG jets. This is achieved through the characterisation of jet at small intervals (12.50–21.74 μs) using high speed Schlieren. Schlieren imaging yields the dynamic changes in the jet structure.

2 Experimental Method

While there are many gasoline injectors available, commercial availability of high pressure gaseous fuel injectors is still in the nascent stage. Thus a multi-hole solenoid based gasoline injector was modified to inject gaseous fuels. Although not an entirely new concept [1–3], design of the sealing arrangement, internal configuration, and a single-hole injector tip gives the most realistic configuration to a production unit.

For the injector selected (Bosch #261,500,056) the 3 mm diameter ball provided a sufficient working boundary for sealing against a 4 mm Ø internal surface and a 1 mm Ø orifice. The curved internal surface was necessitated in prevention of lipping induced from the high impact needle closure, and to provide a large sealing surface area. Final injector design is depicted in Fig. 1 with features that include; an adjustable tip position (4), two Viton seals to for gas containment past the thread (5) and a brass (Grade 385) body.

A constant volume chamber (CVC) with up to four orthogonally positioned 150 mm Ø fused silica (Corning Code 7940) windows were employed. The injector is mounted in a vertical position on top of the CVC with all other services accessed through the bottom (i.e. N_2 purge line). To achieve substantially high pressure ratios the chamber was evacuated and retained at vacuum pressures down to 0.25 bar (abs).

Schlieren HSV affords a non-intrusive line-of-sight investigation which is particularly convenient for resolving the dynamic processes of gaseous fuel delivery behavior. The principal of the Schlieren technique are well documented [4] and is a commonly employed method in this field i.e. [5].

Images were acquired using a folded Z-Type configuration where a continuous Xenon lamp (300 W Xenon lamp Olympus CLV U20) illuminated an imaging field

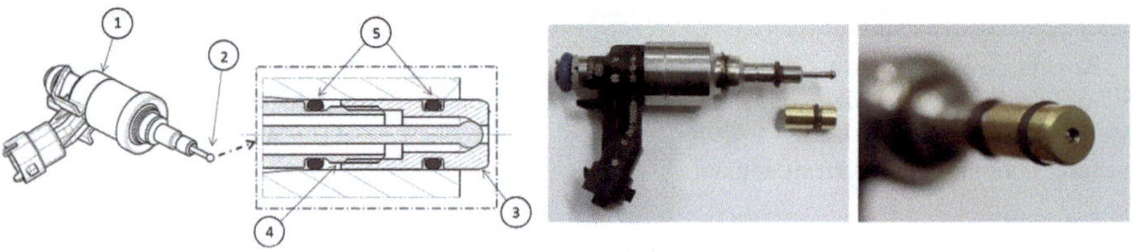

Fig. 1 Injector and tip design features (*left*) and final assembly (*right*)

up to 152 mm Ø utilizing parabolic mirrors with a focal length of 1524 mm. The camera utilized in this work was a Vision Research Phantom v1610 married to either a Sinar Sinaron S 300 mm or a Rodenstock Apo-Ronar 800 mm lens for the freestream and nearfield tests, respectively. This combination resulted in an imaging capacity of 0.157 mm/pixel (384 × 640) at 46,000 fps (21.74 μs between frames) for the freestream conditions. As for the near field scenarios, imaging capacity was performed at a higher resolution of 0.087 mm/pixel (256 × 400) at 80,000 fps (12.50 μs between frames). This provided sufficient resolution to resolve the dynamic properties in adequate scale for the jet nozzle diameter.

An in-house Matlab processing script was developed to extract useful qualitative and quantitative information from each video frame.

Two jet regions of processing were considered; macro and internal. The macro geometric properties are employed for the freestream imaging where upon image segmentation the jet boundary provides the axial and radial penetrations and velocities and the spread angle. Segmentation is based on a change in pixel intensity from a stagnant background image and is bounded by a thresholding edge detection technique. The accuracy of such a technique is estimated at ±3 pixels which yields an uncertainty approximation between 0.26 to 0.47 mm.

The macro and compressible flow processing properties have been outlined at length in [6]. For clarity Fig. 2 depicts the key selection areas for the compressible structure where (1) is the barrel width at the nozzle exit (2) is the triple point location (developing and developed, respectively) and (3) is the reflection angle coordinate.

From these selected coordinates the barrel length and Mach disc diameter are presented. The barrel length is defined as the axial distance from the nozzle exit to the triple-point whilst the Mach disc height is the radial distance between the two triple points where appropriate. This representation gives quantification to the evolution of the classical shock propagation for highly underexpanded jets.

A third set of coordinates were tagged onto the image to represent appropriate positions along the left and right Mach disc reflection lines. The reflection angle is calculated from these locations alongside the previously selected triple points. The reflection angle presented in this work is the average of the two angle calculations.

The compressible flow structures are dominate features in the underexpanded jets considered and have a strong bearing on the succeeding flow and the overall delivery. Hence Mach disc diameter, barrel length and triple point reflection angle measurements contribute to the overall aims of this work.

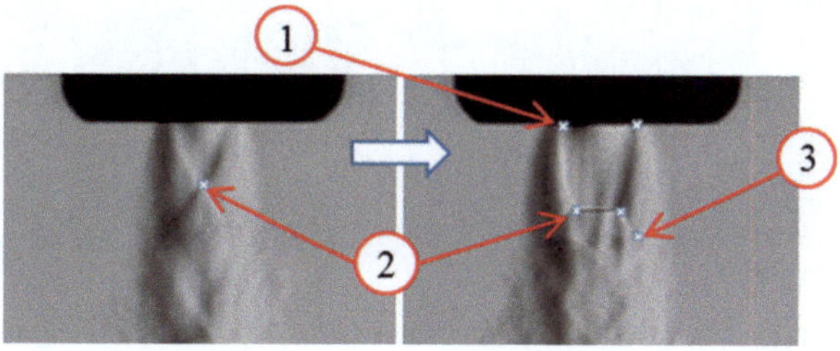

Fig. 2 Compressible flow feature selection locations

3 Experimental Conditions

Jets issued from a nozzle at exit pressure ratios (*NPR*) greater than two are considered to be of highly underexpanded nature [7]. The *NPR* is related to the upstream supply pressure ratio, *PR*, through isentropic considerations utilizing the ratio of specific heats, γ. This approach has been widely incorporated including for conditions close to this work [8, 9]. The use of *NPR* is especially useful when multiple fuels are used.

Several key flow conditions pertaining to both fundamental and practical objectives are experimentally tested. This range of *PR* values are expected to cover the current practical limitations of gaseous DI fuel injection and the near-future trends. The flow is allowed to naturally develop and as such all conditions are considered to be freestream flow. The near-field conditions presented allow for greater resolution for development lengths, $x/d \lesssim 30$. In total seven test positions have been selected for $40 < PR < 320$.

4 Results

Examples of the imaging conditions represented through Fig. 3 and later on in Fig. 7. Figure 3 depicts the variation of jet development between conditions at $PR = 80$ and $PR = 120$. It can observed that in the initial development period, $t < 175$ µs, there is little variation in the jet profile.

4.1 Macro-geometric Structures

The freestream macro properties are analysed within this section to address the goals surrounding fuel targeting and distribution. The axial penetration, presented here normalised with respect to the nozzle diameter, is highly dependent on the

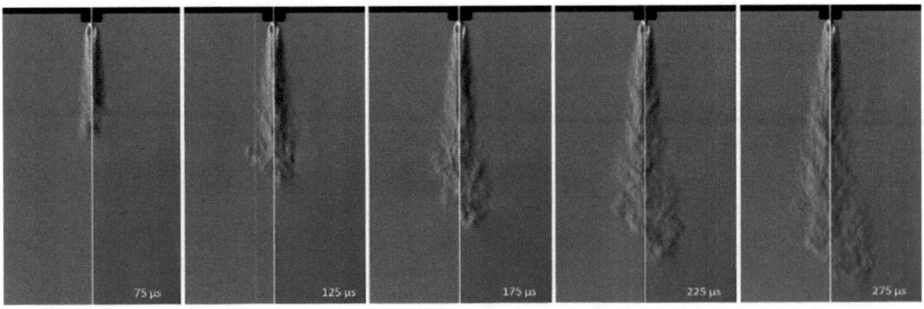

Fig. 3 Back-to-back jet development sequence for PR = 80 (*left*) and PR = 120 (*right*)

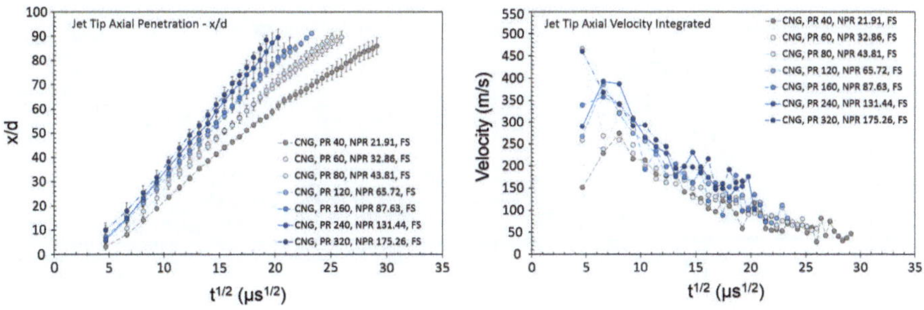

Fig. 4 Jet tip axial penetration (*left*) and velocity (*right*)

interaction of between the two fluids. The penetration rate is a function of the *PR* for the CNG jets where it can be observed in Fig. 4 (left) that the rate of axial development (*x/d*) increases for an increasing *PR*.

Through knowing or being able to predict the axial penetration a greater understanding of the fuel delivery and targeting is achieved. For example, one can predict the time taken for the CNG jet to reach a location of interest like the piston surface.

The axial velocity is, of course, linked to jet tip penetration where similar *PR* dependencies are observed. It can be correlated through the velocity decay profile (Fig. 7) and axial penetration that the CNG jets can penetrate up to x/d = 16 at a velocity up to 760 m/s, for the first image aSOI (4.66 $\mu s^{1/2}$).

It is clear that if the early development of these very high *PR* jets are to be captured then higher resolution (spatial and temporal) imaging is needed focusing on the development up to x/d = 15. Unfortunately due to the horizontal cut-off imaging and thus low macro axial boundary definition, the nearfield imaging data employed in this work is unusable.

It is, however, clear that even at large *x/d* values or large aSOI times, the jet is moving at substantially high velocities. The greater the jet velocity the higher the ambient fluid entrainment rate will be increasing the local mixing. It can also be seen that the jet velocity will decay sharply over the first 400 μs indicating a high rate of fluidic interaction between the CNG jet and nitrogen environment. It should be kept in mind that the momentum of the jet in this region is also increasing at the nozzle and that the fuelling period is several thousand micro seconds (i.e. the jet considered is not of a "puff" in nature).

The radial penetration trends are again in line with the considerations outlined for the axial development. The radial penetration is taken as the maximum radial measurement which typically occurs at a location between 60–90 % of the axial penetration. The exact location is dependent on the interaction between the jet and environment for the impulsively started transient jets considered here. Figure 8 depicts the nozzle normalised maximum radial penetration.

The velocity of the radially developing jet is taken as the total velocity and as such if considering only one side of the jet one must halve the values presented. This is an appropriate estimation due to the highly symmetrical flow; typically

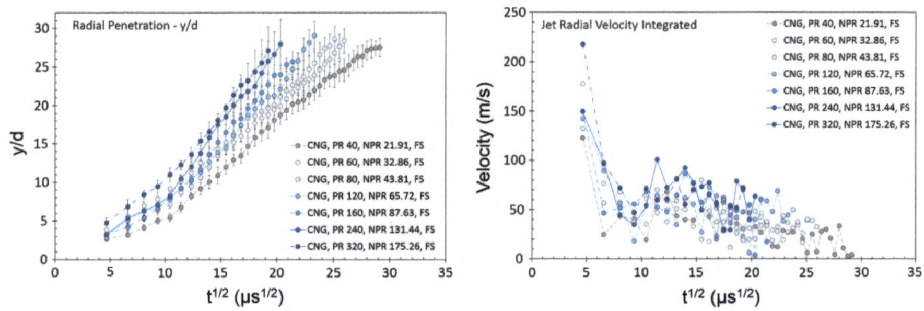

Fig. 5 Radial penetration (*left*) and velocity (*right*)

above 90 %. The radial velocity decays very rapidly where it nears asymptote behavior after $t^{1/2} \gtrsim 11\ \mu s^{1/2}$. The radial development in the main body of the jet is halted through two main reasons: the ambient fluid is entering the jet at normal-angles through the entrainment processes and through the viscous mixing effects at the shear layer (Fig. 5).

If one is to consider stratified charge operation; where the fuel is typically ignited at it's boundary via a spark plug, the radial (and axial) velocity component gives an indication of the likelihood of ignition. The ignition of the fuel or conversely extinguishment of the flame kernel is highly susceptible to the flow velocity across the spark plug.

The spread angle also aids in this sentiment by providing greater clarity on the geometric nature of the jet and its targeting profile.

As introduced in the experimental methods section the spread angle was measured up to various axial locations. The closest location to the nozzle exit is 0.33× and is represented in Fig. 6 (left). This location gives the most realistic results for ascertaining the influence of the nearfield shock behavior but comes at the cost of higher variability in the early stages aSOI, due to the higher turbulence induced from the impulsive starting of the jet. It can be seen that in general the spread angle appears to increase with increasing PR but an absolute understanding is clouded by the measurement variability. All error bars presented in this work are representative of measurement range.

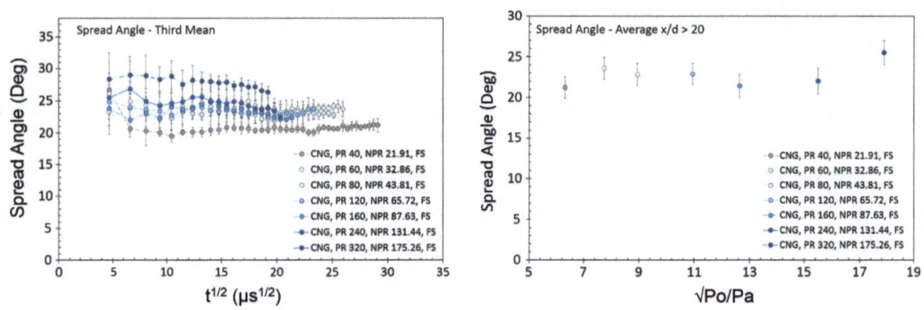

Fig. 6 Spread angle third mean (*left*) and mean for fully developed jets (*right*)

Additionally, all three spread angle methods have been collapsed for jet development lengths: $x/d > 20$, depicted in Fig. 6 (right). It can be seen that the spread angle circulates around 22° with the exception of the $PR = 320$ case. Whilst this appears as an anomaly or an outlier there was a trend in increasing spread angle for CNG jets under identical conditions beyond $PR = 320$ (not included in this current work).

4.2 Shock Structures

Examples of the effects of delivery pressure ratio are shown in Fig. 7 where there are marked differences in the region immediately after the nozzle. Typically, images of this nature are shown for much larger nozzle orifice diameters and even more seldom for transient flows.

The evolution of the Mach disc has rarely been analysed in modern literature [10, 11] with the majority of the status-quo residing in steady state jets. This work provides a highly resolved depiction of the formations of the highly-underexpanded compressible phenomena. Whilst the process of evolution has been detailed to a similar resolution in other works by the current authors this depiction is unique; the *PR* spread is large at 10:1 and the nozzle diameter is small at 1.00 mm Ø.

Interestingly, the formation process was not as predicted. The jet undergoes a process where the growth of the shock structure is normal; increasing with time and nozzle exit pressure stabilization but reaches a maximum at $t \cong 13.22\ \mu s^{1/2}$ after which a decrease or contraction occurs to $t^{1/2} \cong 16.50\ \mu s^{1/2}$ followed by a slow increase and then a largely stable period. This process is seen for all experimental cases where an increase in magnitude is proportional to the *PR*. This affect coincides near-identically with the needle lift (not shown).

The average Mach disc diameter has been plotted against individual $\sqrt{PQ}$ values as depicted in Fig. 8. The high linearity ($R^2 = 99.29\ \%$) of the Mach disc diameter indicates that there is a strong agreement that the transient and steady state condition of the Mach disc diameter is dependent of both the *PR* and the conditions at which the jet exists; needle lift profile and internal injector geometry.

The barrel length of the underexpanded shock is a marker for the location of where the standing Mach disc shock occurs from the nozzle exit. This aspect of the compressible flow feature has/is widely studied and used also to correlate

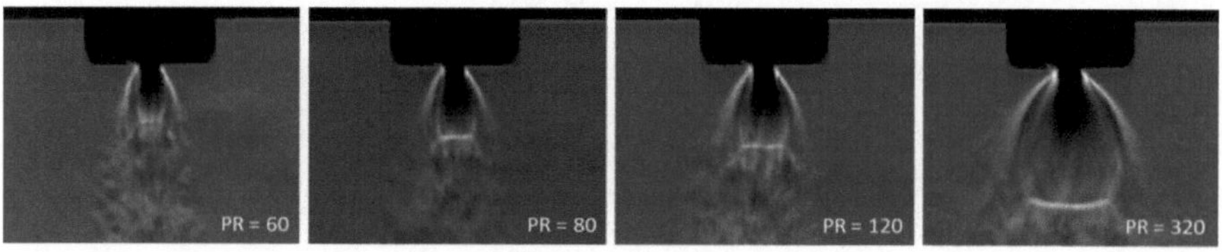

Fig. 7 Nearfield shock structures for $t^{1/2} = 26.7\ \mu s^{1/2}$ aSOI

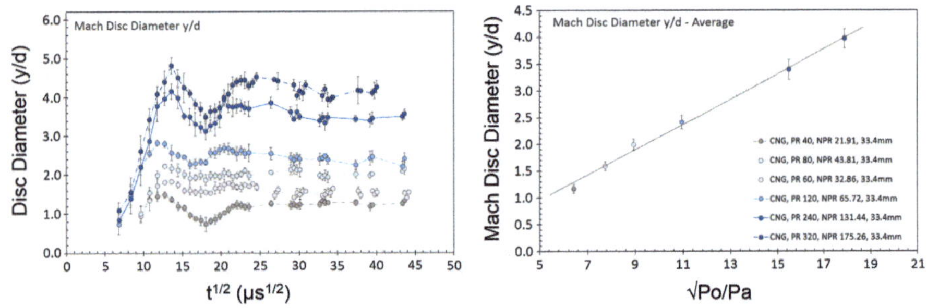

Fig. 8 Mach disc diameter evolution (*left*) mean diameter at steady state (*right*)

computational models [12]. Through knowing the distance of the barrel in conjunction with the Mach disc diameter (and reflection angle) a fairly accurate depiction of nearfield conditions can be acquired; useful in accurately positioning supersonic and subsonic flow velocities and for spatial (targeting) information.

The evolution of the barrel length is presented in Fig. 9 (left) where it can be observed that a similar trend to the Mach disc diameter is present. Collapsing the data locations to average $\sqrt{PQ}$ normalised values, Fig. 9 (right) depicts the dependence on the supply to ambient pressure ratio. As with the Mach disc diameter, the barrel length also proved to show a high linear dependency on the $\sqrt{PQ}$.

The shock reflection angle (α) development trend is aligned with the other shock structures where an initial growth, overshoot and settling period occurs. Moreover, there is a strong oscillation period observed even for the otherwise steady shock structures. The average oscillation of the reflection angle after the initial transition period occurs between 25 and 40° and is similar to expectations derived from lower PR CNG jets (27° < α < 39°) and other recent works or similar nature [13]. The reflection angle development is depicted in Fig. 10 (left). The mean reflection angle for all pressure ratios examined, Fig. 10 (right), indicates that the reflection angle is not dependent on the magnitude of the shock phenomena. Excluding PR = 60 ($\sqrt{PR} = 7.746$) the mean reflection angle is approximately 30.32°. This figure is again consistent with expectations from previous work on substantially lower PR jets.

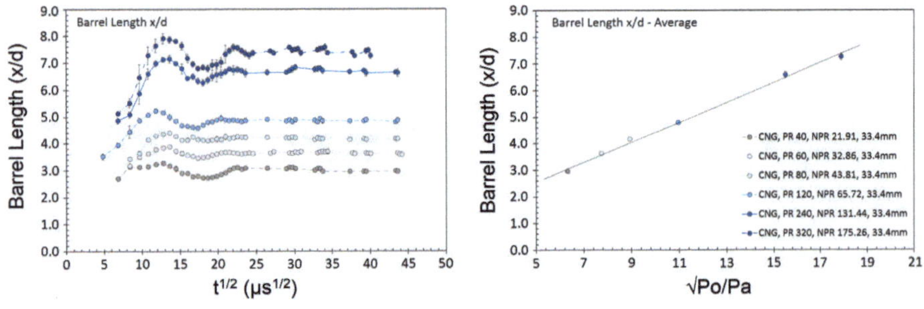

Fig. 9 Barrel length evolution (*left*) and mean length at steady state (*right*)

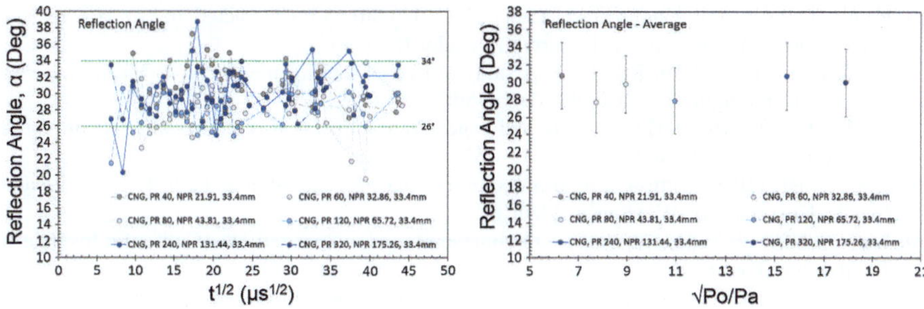

Fig. 10 Reflection angle (*left*) and the mean across the steady region (*right*)

5 Conclusion

CNG jets issued at conditions relevant to DI operation have been studied with high temporal and spatial resolution which allowed clear quantification of the macro and internal structures and the interplay between the two. A brief summary of the key conclusions are as follows:

1. A unique method of adapting a gasoline injector for use with CNG has been proposed and utilized in a high supply pressure environment and significantly large pressure ratios.
2. High resolution characterization of the growth time for both Mach barrel length and Mach disc diameter. The temporal evolution is independent of PR.
3. The Mach disc diameter and its location; the barrel length, are shown to be highly predicable across the range tested within a linear fit.
4. The Mach disc reflection angle has been characterized at high resolution, where it is shown that even before the Mach disc has fully developed the reflection angle remain constant within its oscillation period.

References

1. Bovo M, Rojo B (2013) Single pulse jet impingement on inclined surface, heat transfer and flow field. SAE Int
2. Johansen LCR, de Benito Sienes E, Dahlander P (2013) Analysis of transient compressible gas jets using high speed schlieren imaging. SAE Int
3. Baert R, Klaassen A, Doosje E (2010) Direct injection of high pressure gas: scaling properties of pulsed turbulent jets. SAE Int J Engines 3(2):383–395
4. Settles GS (2001) Schlieren and shadowgraph techniques: visualizing phenomena in transparent media, 1st edn. Springer, Berlin, 376 pp
5. Hajialimohammadi A et al (2013) Time resolved characteristics of gaseous jet injected by a group-hole nozzle. Fuel 113:497–505

6. Rogers T, Petersen P, Koopmans L, Lappas P (2014) Structural characteristics of gaseous fuels Part I: Port fuel injection. Fuel (Currently Under Review)
7. Donaldson Cd, Snedeker RS (1971) A study of free jet impingement. Part 1. Mean properties of free and impinging jets. J Fluid Mech 45(02):281–319
8. Birch AD et al (1984) The structure and concentration decay of high pressure jets of natural gas. Combust Sci Technol 36(5–6):249–261
9. Petersen BR (2006) Transient high-pressure hydrogen jet measurements, in Engine Research Center. University of Winconsin, Madison
10. Lacerda NL (1987) On the start up of supersonic underexpanded jets. California Institute of Technology, Pasadena
11. Maté B et al (2001) Experimental and numerical investigation of an axisymmetric supersonic jet. J Fluid Mech 426:177–197
12. Otobe Y et al (2008) Influence of nozzle geometry on the near-field structure of a highly underexpanded sonic jet. J Fluids Struct 24(2):281–293
13. Yu J et al (2013) Visualization and analysis of the characteristics of transitional underexpanded jets. Int J Heat Fluid Flow 44:140–154

Development of a Methodology to Evaluate the Energy Efficiency of Heavy Equipment

P. Scherer and M. Geimer

Abstract It was investigated how the energy efficiency of heavy equipment can be evaluated. The wide range of applicability of heavy equipment, as well as the existence of one or more power take-offs requires a special approach for evaluating the energy efficiency of these machines. Therefore, different types of heavy equipment vehicles were equipped with sensors and measurement instrumentation throughout the whole powertrain. With these vehicles, field tests were realized in order to receive perception about the application area and movement- and load spectra of these vehicles during the process of duty cycle operations. Parameters were varied in order to receive information about the impact of environmental effects on the movement- and load spectra. For the analysis of these measurements a software package was developed, which supports the analysis process of measurement data. As a result of the analysis, different standardized field test cycles were designed exemplarily. The driven test cycles were finally object to the evaluation of the energy efficiency. Based on this strategy, a methodology was developed, which provides a uniform way to evaluate the energy efficiency of the analyzed vehicles. The concept is valid for all types of heavy equipment. It is foreseeable that a realization of this methodology with additional vehicles will likewise deliver an energy efficiency evaluation which corresponds to the predicted application area.

P. Scherer (✉) · M. Geimer
Chair of Mobile Machines (Mobima), Karlsruhe Institute of Technology (KIT), Rintheimer Querallee 2, 76131 Karlsruhe, Germany
e-mail: philipp.scherer@kit.edu

M. Geimer
e-mail: marcus.geimer@kit.edu

I. Denbratt et al. (eds.), *Sustainable Automotive Technologies 2014*,
Lecture Notes in Mobility, DOI 10.1007/978-3-319-17999-5_11

1 Introduction

The energy efficiency of vehicles is usually measured in standardized test cycles, which are designed for a class of vehicles. The range of use cases and the variety and flexibility of operating modes of a class of vehicles raise the complexity for the design of precise standardized test cycles. Vehicles with single use cases, e.g. a car, are comparatively easy to test, considering aspects of energy efficiency. For heavy equipment vehicles, there is usually not only one purpose and application area. A tractor for example can be equipped with different implements and therefore be used for different purposes. Examples for known representatives of existing standardized test cycles for heavy equipment vehicles are:

- Wheel loader: Load and Carry cycles [1]
- Tractor: DLG-PowerMix [2], Nebraska Tractor test [3]
- Forklift truck: Performance test cycle [4]

These test cycles are suited as possibilities to evaluate the energy efficiency of the examined vehicles. Although these test cycles represent appropriate testing methods for the regarded vehicles, no validated methodology or database for the derivation of the applied test cycle setups is publicly documented. Furthermore, there is a European Union legislative demand for a reduction of greenhouse gases, which affects the automotive industry and to a greater extent also more sectors in the field of off-road vehicles. Strategies for the determination of realistic fuel consumption and carbon dioxide exhaust amounts are needed. Under the project name TEAM [5], a 3 year project has been started, which is focused on the development and integration of new energy efficient propulsion and steering systems for heavy equipment. One task within the project is finding an appropriate way of evaluating the energy efficiency of the enhanced machines, after the powertrain modifications have been applied to it. For various types of agricultural- and construction vehicles, a uniform methodology was developed, in order to design testing methods which meet certain required criteria in the appropriate relation.

2 Challenge of Choosing the Right Testing Method

Dealing with an energy efficiency evaluation methodology, the first aspect to consider is a common comprehension of energy efficiency. As energy efficiency has no uniform global definition [6], it is to mention that the developed methodology in this paper works for a result related efficiency definition. The input parameter in this definition is the total amount of energy E_{in} provided by the energy source (e.g. the energy content per volume unit diesel fuel) of the vehicle, that is consumed during a duty cycle operation. The output is defined as the result R which is generated through the vehicle operation. Regarding construction- and agricultural vehicles, examples for this result would be moved bulk material by a wheel loader when operating a load and carry cycle or a hole with defined dimensions as a result of a drilling process by a drilling rig. The efficiency η_{result} therefore can be defined as:

$$\eta_{result} = \frac{R}{E_{in}} \tag{1}$$

Other existing definitions, such as economic definitions [7] or the energy conversion efficiency definition, are not explicitly part of this methodology.

In order to be able to evaluate the mentioned energy efficiency, the mayor prerequisite is a suitable testing method for the considered class of vehicles. Testing methods are usually realized in the form of standardized test cycles. The test cycles have to match the following criteria:

1. Representing realistic use cases of the vehicle
2. Considering frequently appearing use cases
3. Being highly reproducible
4. Being realizable in an economic way
5. Being as much as possible a real reproduction of the vehicle utilization and its environment

The problem of choosing an adequate testing method has to be solved in order to match the mentioned criteria in an appropriate relation. For heavy equipment vehicles, there is usually not only one purpose and application area. A tractor for example can be equipped with different implements and can therefore be used for different purposes. It is important to notice, that there can be various appropriate testing methods for a certain type of vehicle. Different users of the methodology, e.g. legislators, manufacturers of a vehicle or a certain group of costumers will generate different test cycles, which cover the specific application area of their purposes.

3 Workflow of the Methodology

In order to design test cycles for any type of heavy equipment vehicle, a methodology was designed. This methodology follows various chronological steps, designated as the workflow (Fig. 1). For these steps, methodical tools were created. Some of these methods were made executable through software applications, which were implemented to the methodology. The methodology is characterized by the following features:

- A chronological order of processing steps, designated as the workflow
- A specific terminology, decision guidance documents and mathematical algorithms, designated as methodical tools
- Software tools in order to make the workflow executable

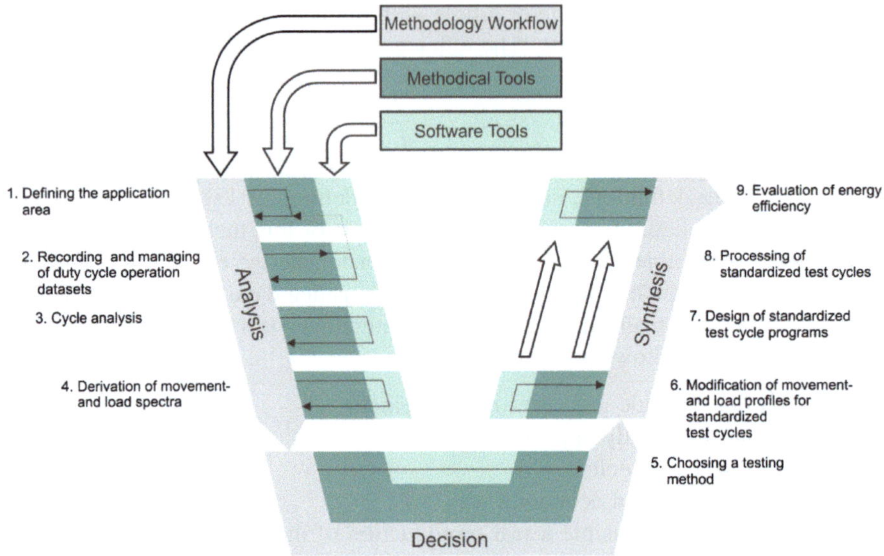

Fig. 1 Workflow of a methodology for the evaluation of heavy equipment vehicles (*Source* Scherer)

The workflow of the methodology can be divided in three sections. The first section is called the section of analysis. Recorded machine operation datasets are analyzed in consecutive steps in order to make proposals for input parameters and associated movement- and load spectra of the vehicles for standardized test cycles. In the section of decision, the generated information in the first section is transferred to a group of experts. The task of this group is to decide and to select, what type of testing method the most suitable for the given circumstances is. This is being done by considering the 5 criteria, which are mentioned in the previous chapter. The section of synthesis concretizes the selected testing methods and leads to the required parameters for the standardized test cycles. Standardized test cycles can be designed and executed with different competing vehicles. The energy efficiency evaluation is the last step of this section.

4 Section of Analysis

Defining an application area: The first step is the definition of an application area for which the evaluation of the selected vehicles is valid. An application area is defined by the occurring use cases and the associated time shares, that a vehicle is facing during a certain time period (Fig. 2). The time shares of each use case are defined by the total amount of considered duty cycle operation periods. The application area can either be assigned by the user, be recorded by a telematics

Fig. 2 Distribution of duty cycle time shares in an application area (*Sources* Scherer, AGCO Fendt)

system [8, 9] or be calculated by the mentioned developed software tool. This calculation is usually based on sensor data from previously recorded datasets. As different applications cause different energy flows through the powertrain of the vehicles, this step is essential to ensure the consideration of appropriate and frequently appearing use cases.

Recording and managing duty cycle operation datasets: The machine operation datasets which are processed in the analysis section derive from various duty cycle operations of the examined vehicles. The content of these datasets are time sequences of sensor data, such as GPS-, implemented CAN-, and additional sensors. The recorded data is classified according to its further usage in the analysis section (Fig. 3):

- Type 1 data provides information about the type of use case, which is processed during a duty cycle operation. This data can be used for the definition of the application area.
- Type 2 data is used to detect single operation steps in a duty cycle operation.
- Type 3 data leads to the derivation of movement- and load spectra.

A software package, which supports the proper processing of the data analysis for the recorded measurement data, was created. This open source software provides functions for the relevant steps of the methodology. For managing the relevant duty cycle operation datasets and the additional data, which is generated via its tool functions, a relational database management system was appended to the software.

Cycle Analysis: Regarding energy efficiency questions, meaningful standardized test cycles should always represent average use cases for a given set of parameters.

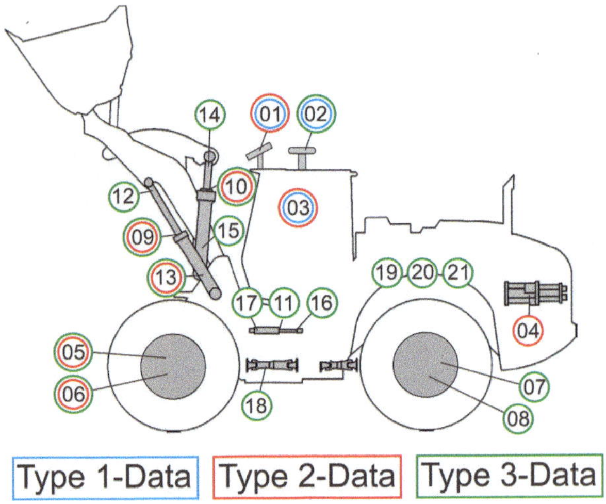

Fig. 3 Measurement equipment of a test vehicle type wheel loader Liebherr L 509 Stereo. Positions: *01* video record, *02* GPS, *03* driver documentation, *04* fuel consumption, *05–08* wheel speed, *09–11* hydraulic cylinder position, *12–17* hydraulic pressure, *18* front axle torque, *19–20* hydraulic pressure of main circuit, *21* volumetric flow of main circuit (*Sources* Scherer, Deiters)

A set of parameters is defined as the sum of variable, mostly environmental parameters, that appear during duty cycle operation. The driver skills, the driving surface and the handling material are examples for parameters, which usually have a big impact on the energy consumption of a vehicle. The idea of cycle analysis is to extract parameter dependent time shares and energy consumptions of machine operation steps from conventional duty cycle datasets (Fig. 4). This extraction is the base for a reassembling of standardized synthetic test cycles in the synthesis section. Continuously repeating operation steps, which are necessary to produce a desired result, are designated as the operation process. Not all operation steps, which appear in conventional recorded duty cycles, are immediately necessary to produce a result. Nevertheless, some of these steps are indispensable parts of a conventional duty cycle, such as e.g. idle mode periods or additional driving sections and are therefore included in the design of standardized test cycles in the synthesis section. In order to be able to make a subdivision into different operation steps, several kinematic movements of the vehicles are extracted from machine operation sensor data. Recorded data of type 2 sensors shows specific characteristics for each operation step. This effect was used to make a reliable subdivision of the different operation steps. From the recorded machines, duty cycle datasets of different use cases are extracted. The cycles are analyzed with regard to the duration of the relevant operation steps. Robust algorithms were implemented to the software package in order to extract the operation processes from the datasets.

Derivation of movement- and load spectra: The design of standardized test cycle programs for a test facility in the synthesis section is based on movement- and load spectra of the investigated vehicles. These standardized test cycles represent

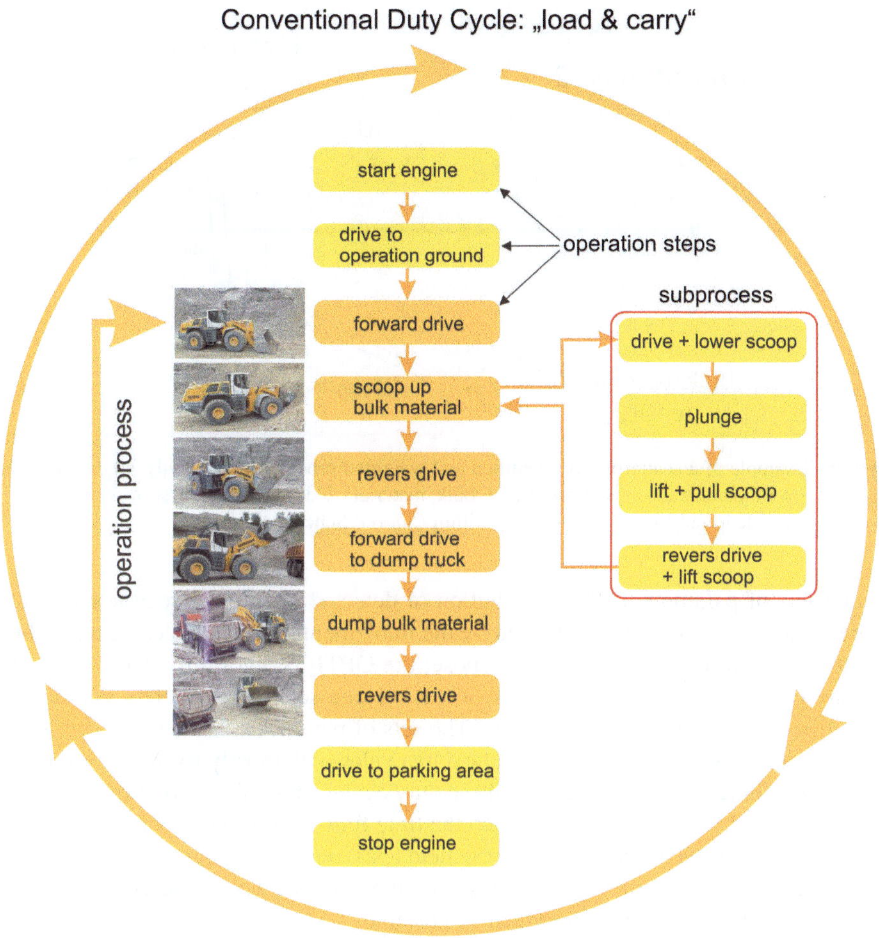

Fig. 4 Operation steps of a conventional load and carry duty cycle (*Source* Scherer)

average duty cycles of these vehicles. Movement spectra are the sum of the position-/time functions of type 3 datasets. These functions are defined as the vehicle tracks in a fixed course on a 2D or 3D-surface. Load spectra are the sum of the flow variable-/time functions and the potential variable-/time functions, that are appearing at the vehicle power output interfaces to its environment during the processing of duty cycles. Examples for flow variables are velocity, speed, electric current and volumetric flow. Technical variables like force, torque, voltage and hydraulic pressure are defined as potential variables. At first, frequently appearing sets of parameters of the recorded type 2 data must be extracted from the recorded datasets, as the derivation of load spectra from type 3 data is only important for frequently and therefore averaged appearing sets of parameters. Detected operation processes within the analyzed type 2 datasets are grouped and added to other processes of the

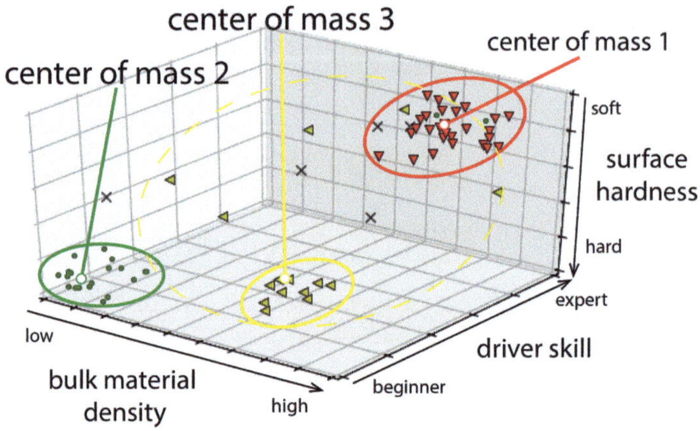

Fig. 5 Example of a scatterplot of identified and weighted processes from analyzed datasets. The plotted set of parameters is defined by: 1 bulk material, 2 driver skills, 3 surface hardness. 3 Clusters are identified by the OPTICS algorithm (*Source* Scherer)

same sets of parameters. The distribution of detected operation processes with different sets of parameters is clustered by the implementation of clustering algorithms to the software package. For this purpose, the OPTICS algorithm [10] shows adequate results. Centers, which represent the average sets of parameters within each identified cluster, are formed (Fig. 5). The sets of parameters which are represented by these centers form the setup for new duty cycles with mostly fixed environmental conditions in test fields. Processing these duty cycles under fixed conditions delivers the mentioned type 3 data, from which the time functions of the spectra are derived. During the duty cycle operation in these field tests, one has to put emphasis on a high periodical and energetic reproducibility of the driven tracks. Therefore, the duty cycles have to be driven as identical as possible.

5 Section of Decision

The section of decision is dedicated to the choice of a convenient testing method. As already mentioned before, it is not possible to designate a test cycle, which fulfills all criteria at most. Therefore, the first step of this section is the election of experts. The task of these experts is to find a reasonable compromise for the selection of a testing method. This decision is based on the compliance of the five criteria which are mentioned in the introduction.

Testing methods are categorized into three categories:

- field tests
- test rig tests
- simulations

Most adequate test facilities represent one of these testing methods or a hybrid of these methods. A defined percentage of each mentioned criterion is covered by the test programs, which are implemented to a test facility. The experts, with regard to the demands of the user, specify the minimal acceptable degree of fulfillment for each criterion. With this input information, those testing methods and test facilities can be excluded, which do not fulfill the specifications. The main task of the experts is the choice of the most convenient testing method afterwards. The covering of the first two criteria (frequently and realistic use cases) is ensured by the processing of the previous analysis section. The remaining criteria are described as follows:

Economic feasibility: A testing method and the appearing costs for it must be economically justifiable. These costs can be subdivided into fixed- and variable costs. Depending on the testing method, the total resulting costs (e.g. daily rates for the facility use, salaries for the drivers, wear and tear parts) for the user that appear when processing standardized test cycles in a test facility have to be considered as a criterion.

Reproducibility: The determination of reproducibility of a testing method is an indispensable need for the energy efficiency evaluation. The less reproducibility a testing method has, the less explanatory power does it have for the evaluation. Reproducibility was defined as a deviation of time duration and energy consumption of a measured operation process to the average operation process. For field test cycles, this deviation is determined from selected type 3 datasets. For test rig measurements, this information can be gathered from test cycle measurements in the test rig. Simulations usually are supposed to have a reproducibility of up to 100 %.

Reproduction of Reality: Test cycles represent reproductions of real duty cycles. It is assumed that field test cycles show the biggest reproduction rate of real use cases with up to 100 %, being followed by test rig tests and simulations. Pure simulation test cycles are considered as a 0 % reproduction rate of reality. Every parameter, for example the deployment of a real driver or the use of an implement, a tire-to soil contact or the usage of a handling material in the testing method contribute percentages to the reproduction of reality.

The task of the elected experts is to collect the necessary information concerning the criteria of relevant testing methods. With the help of a spreadsheet application, the percentages of every criterion for every testing method are calculated and the advantages are weighted. After the decision for a testing method the relevant movement- and load spectra are transferred to the first step of the synthesis section.

6 Section of Synthesis

The sources of the derived spectra are as mentioned before recorded type 3 datasets. As in most cases, sensors cannot be located at the exact vehicle power output interfaces to its environment (see sensor positions in Fig. 3), e.g. the wheel/soil, or implement/handling material interfaces, of a vehicle, the derived movement- and load spectra are modified in step 6 in order to apply them as forced input spectra at

Model	Total	
	Ranking	%-Top Runner
Manufacturer 1 Model A	1	103%
Top Runner	***2***	***100%***
Manufacturer 3 Model C	3	97,7%
Manufacturer 4 Model D	4	94,7%

Front Loading		
Ranking	Consumption	%-Top Runner
1	4,9 l	106,1%
3 (TR-Front Loading)	***5,2 l***	***100,0%***
2	5,0 l	104,0%
4	5,5 l	94,5%

Baling		
Ranking	Consumption	%-Top Runner
2	15,1 l	98,0%
1 (TR-Baling)	***14,8 l***	***100%***
3	15,4 l	96,1%
4	15,9 l	93,1%

Plowing		
Ranking	Consumption	%-Top Runner
1	7,8 l	103,8%
2 (TR-Plowing)	***8,1 l***	***100,0%***
4	8,7 l	93,1%
3	8,4 l	96,4%

Fig. 6 Possible results of standardized test cycles of tractors in a specific engine-power class (Source Scherer)

the interfaces from the vehicle to the test facility. The complete design of standardized test cycles in a test facility is arranged by the staff of the facility in step 7. The standardized test cycles include the modified spectra and must be made executable with all vehicles of the same type and engine-power class. If a simulation was chosen as testing method, this step includes tasks like the creation of the simulation environment, the implementation of models and the modified spectra, as well as the choice of the hardware. After the processing of the standardized test cycles with the selected vehicles, the evaluation step follows. As mentioned before, the evaluation is based on the ratio of a Result R in relation to the consumed energy E_{in} of the vehicle. This ratio is formed by the measurements of energy consumption during the processing of the standardized test cycles in relation to the generated result. In a field test, this result can usually be identified visually, for example by the moved amount of bulk material or the plowed surface of a field in relation to its energy consumption. In a test rig test cycle, this ratio is identified by the amount of energy, which is consumed during the processing of the applied standardized test cycle program. The efficiency evaluation in step 9 is made comparatively between the tested vehicles. Similarly to the Top Runner approach [11], the vehicle with the best results in the test cycle is designated as the Top-Runner. The efficiency evaluation of the Top Runner is defined as 100 % efficient. Efficiency ratios of other tested vehicles are indicated as a percentage of the achieved ratio by the Top Runner. If more than one use case is covered by the standardized test cycles, the overall Top Runner does not necessarily have the best results at every covered use case (Fig. 6).

7 Conclusion/Summary

In this article a new methodology to evaluate the energy efficiency of heavy equipment vehicles was proposed. This methodology is valid for the most kinds of off-road vehicles, which possess a powertrain with additional work functions. The advantage of the methodology is its flexibility, considering different users and application areas of the vehicles. The processing of the methodology workflow is made realizable through the attached software programs and a database system. The evaluation of the energy efficiency of the vehicles is finally made as a comparative result between different vehicles. It is to mention, that depending on various factors, e.g. the application area of the tested vehicles or the chosen testing method different results for the energy efficiency for the vehicles will appear.

References

1. Deiters H (2009) Standardisierung von Lastzyklen zur Beurteilung der Effizienz mobiler Arbeitsmaschinen, Shaker
2. Degrell O, Feuerstein T (1999) DLG-PowerMixTM ein praxisorientierter Traktorentest. In: VDI-Berichte(1798)
3. Grisso R, Pitman R et al (2001) Gear up and throttle down to save fuel. Virginia cooperative extension paper no 442–450
4. VDI-Gesellschaft Produktion und Logistik (GPL) (2012) Typenblätter für Flurfoerderzeuge. In: VDI Richtlinie 2198, Beuth Verlag Gmbh
5. Weber J, Scherer P et al (2013) TEAM - Effizienz treibt uns an, http://www.team-mobile maschinen.de. Accessed 07 April 2014
6. Pehnt M (2010) Energieeffzienz. Springer, Berlin. doi:10.1007/978-3-642-14251-2_1
7. Irrek W, Thomas S (2008) Definition Energieeffizienz. In: Wuppertal Institut für Klima, Umwelt, Energie GmbH. http://wupperinst.org/uploads/tx_wupperinst/energieeffizienz_definition.pdf. Accessed 07 April 2014
8. Ryken M, L'heureux (2013) Understanding customer duty cycle through JDLinkTM. In: 71st international conference on agricultural engineering. VDI Verlag GmbH
9. von Hyningen-Huene M, Baldinger M, Pöttinger A (2010) Tractor-implement-automation and it's application to a tractor-loader wagon combination. In: 2nd international conference on machine control and guidance, Schriftenreihe des Instituts für Geodäsie und Geoinformation, pp 171–185
10. Ankerst M, Breunig M, Kriegel H-P (1999) OPTICS: ordering points to identify the clustering structure. In: ACM SIGMOD international conference on management of data. ACM Press, pp 49–60
11. Kimura O (2010) Japanese top-runner approach for energy efficiency standards. In: Session paper: SERC09035.CRIEPI

Part IV
Sustainability

Full Cost Accounting in the Automotive Industry: A Systematic Review and Methodology Proposal

D. Jasinski, J. Meredith and K. Kirwan

Abstract To date, it is not known how Full Cost Accounting (FCA) can be applied in the automotive context. Therefore the objective of this paper is to review existing methodological approaches in FCA studies and identify the most appropriate for the automotive sector. This systematic literature review of FCA studies extracted ten different methodological approaches in this field. The Sustainability Assessment Model (SAM) is the most developed FCA methodology applicable to automotive organisations. Its strengths are the ability to provide both physical and monetary metrics for sustainability assessment, its flexibility and the ability to combine multiple sustainability dimensions. The SAM takes the full life cycle approach for the sustainability assessment which allows managers to make the best possible choices at the design stage with knowledge about all possible economic, social and environmental costs of the vehicle throughout its lifetime.

1 Introduction

The car industry is fundamental to the economy of many countries (for example, Germany, Japan, Korea, USA, Italy and France). The British Government announced that the UK automotive industry accounts for 5.2 % of manufacturing employment and 7.3 % of manufacturing output with nearly £60 billion turnover in 2012 [6]. However, this tremendous economic performance has significant consequence for the environment and society such as: depletion of natural resource, contribution to the global warming, acidification of the atmosphere, congestion, accidents and noise. Therefore, the automotive sector has increasingly been under pressure from policymakers and other stakeholders to improve sustainability performance.

D. Jasinski (✉) · K. Kirwan
University of Warwick, Coventry, UK
e-mail: D.Jasinski@warwick.ac.uk

J. Meredith
University of Sheffield, Sheffield, UK

I. Denbratt et al. (eds.), *Sustainable Automotive Technologies 2014*,
Lecture Notes in Mobility, DOI 10.1007/978-3-319-17999-5_12

The examples of strategies used by automotive organizations to mitigate social and environmental effects include investment in clean technologies, design for sustainability and creating value for local and global communities. In order to manage these strategies economically, there is an increasing demand for robust decision-making tools that measure and inform managers about economic, environmental and social consequences of their decisions. This study proposes the Full Cost Accounting (FCA) concept as a practical tool to deal with the complexity of triple bottom line decisions in the automotive environment. It embraces both internal and external sustainability impacts and translates them into the widely known and accepted business language of 'money'.

This systematic literature review aims to answer two research questions: (1) What are the methodological implications of past experiences with FCA? and (2) What is the best methodological approach for applying FCA in the automotive context? The review has identified the methodological streams which have been developed to date and selects the Sustainability Assessment Model (SAM) as the one that fits the specifications and needs of an automotive business.

2 Motivation for FCA Tools

Road transport contributes one-fifth of the European Union's total emissions of carbon dioxide in which cars are responsible for approximately 12 % [7]. Due to the high ecological and social footprint of the automotive sector, car manufacturers have increasingly been under pressure from policymakers and other stakeholders to improve the sustainability performance of vehicles at every stage of the life cycle. For example, the End-of-Life Vehicle (ELV) Directive requires automotive companies to think more broadly about future costs (for instance, recycling, disassembly, disposal costs) from current production. Euro 5 and 6 emission limits and the European Environmental Noise Directive place liability on car manufacturers for negative effects from the use of vehicles. Furthermore, British automotive organisations are obliged to improve the sustainability performance at their manufacturing sites based on the Carbon Reduction Commitment (CRC) Energy Efficiency Scheme, Climate Change Levy (CCL) tax, landfill tax and health and safety regulations. All these legal requirements put the responsibility on car manufacturers to improve the social and environmental impacts of their products at every stage of their life cycle.

In order to make the best possible choices, managers and design engineers need effective and credible measurement tools to understand all of the economic, social and environmental impacts of their decisions as early in the product design cycle as possible [8]. Vehicle design and development is considered the most important stage of the automobile life cycle because it determines the lifetime costs and the overall sustainability performance such as fuel consumption, materials composition, safety and emissions [11]. The decisions made at this point have economic, environmental and social implications throughout the entire lifetime of the vehicle.

Adopting FCA tools at the design stage assists in planning for sustainable automobiles by evaluating alternative options (for example, selecting material or processes) and systematically considering the environmental, economic and social performance during the process of developing a vehicle.

2.1 Decision-making Reasons

Life Cycle Assessment (LCA) is a widely accepted technique to assess the environmental performance of a vehicle from the cradle to the grave (from extraction of the raw materials to its final disposal). LCA techniques provide physical information about the internal and external environmental impacts of automobiles in different units such as tonnes of carbon dioxide, sulphur dioxide, nitrogen oxides, cubic metres of water or megawatts of energy. However, different management levels and departments require different sustainability information (Fig. 1). Top management primarily understand the language of money therefore they require monetary and aggregated data to make strategic decisions. Middle management must be bilingual as they need to convert physical information generated by LCA to monetary terms for top management and vice versa to supply lower management and engineers with physical information so that they can understand the implications of strategic decisions made by senior management. Thus, FCA supports strategic choices and improves the communication between different management levels by turning sustainability into the most understandable and widely accepted business language of 'money' [2].

2.2 Business Reasons

A measurement system based on FCA can expose new business or investment opportunities for automotive organisations by measuring internal and external sustainability costs and benefits. Car manufacturing is resource and energy-intensive, representing significant cost of a car and exposes a company to additional costs such

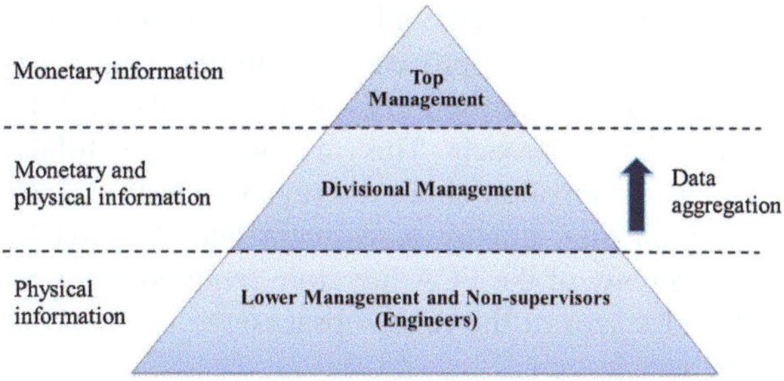

Fig. 1 Management groups and sustainability information demand (adapted from [4])

as 'green' taxes, penalties and fines. Hence, measuring internal impacts may provide immediate financial gains in the form of lower costs from reduced waste sent to landfill, water and energy consumption or carbon emissions. For example, Toyota saved approximately 38 billion yen (equivalent to half a billion US dollars) between 2008 and 2010 mainly by reducing energy consumption, reducing waste processing costs, selling recyclable goods and utilizing other environmentally friendly technologies (Zokaei et al. [16]). As with carbon dioxide, externalities can be internalised at a certain point in time therefore they are considered as future costs. Knowing and anticipating them before they arise can assist in a company's strategic planning and risk management.

2.3 Practical Reasons

Sustainability is a complex, multi-disciplinary issue and requires a multi-disciplinary approach with engagement from various professions: accountants, environmental economists, engineers, human resources, etc. Senior management in automotive organisations receives a variety of information from different departments to make sustainable decisions regarding products, processes and the whole organisation. FCA's ability to combine the multi-disciplinary issues of sustainability into a single framework is an effective way of communicating trade-offs and outcomes from the assessment [2]. Having all these elements together facilitates the decision-making process by providing a more transparent and clear picture about the sustainability performance of a product, process or the whole organisation.

Although there is a case for applying FCA in the automotive business, it is still unknown which FCA approach would best fit. Past experiments with FCA have provided different methodological approaches in this field. The following section explores these methodologies by systematically reviewing the available FCA literature.

3 Methods for the Systematic Review

A comprehensive and systematic review of the available studies on FCA is needed to identify all the methodological streams that have been developed to date. The advantages of this approach over the conventional review method are objectivity, transparency, minimised risk of bias in the results, and its methodological and standardised approach. This review process followed the review protocol that contains information about the review question, inclusion criteria, search strategy, data extraction, quality assessment and data synthesis (Table 1). Diversity and heterogeneity of the FCA literature required a combination of different methods at different stages of the review process.

Table 1 Review protocol

Step	Research question/methods
Review question	What are the methodological implications of past experiences with FCA?
Inclusion criteria	**Specific group**: All FCA studies
	Intervention: Any studies that examine past and current experience with FCA
	Outcome: All studies that represent, constitute or strengthen the methodological approach
Exclusion criteria	Presentations, book reviews, comments and all studies reported in non-English language
Searching the literature	**Methods**: Database searching, grey literature searching, reference list checking, citation searching and consultation with an expert
	Databases searched: Google Scholar, ScienceDirect, Emerald Insight, Wiley Online, Web of Science
	Keywords for database searching: 'Full cost accounting', 'total cost accounting', 'full environmental cost accounting', 'total impact accounting' and a combination of the following terms: 'accounting', 'valuing', 'externalities', 'external cost', 'social accounts' and 'environmental accounts'
Quality assessment	**Methods**: Hierarchy of study design (experimental, observational, expert opinion) and quality checklist (lists of questions appropriate to the research question)
Data extraction	**Categories developed from relevant FCA studies**: Place of study, type of industry, type of focus, type of assessment, scope of the assessment, type of information measured and the valuation method
	Software used for extracting data: Microsoft access
Data synthesis	**Methods**: Narrative synthesis, developed categories from a detailed examination of all FCA studies
	Presentation methods: Tables, graphs, matrices and qualitative thematic analysis

The process of selecting FCA studies based on the review protocol is presented in Fig. 2. Forty six FCA studies were selected for the review process with 35 empirical (experiments and case studies) and conceptual FCA applications.

4 Results and Discussion

The outcome of past experimentation with FCA is the development of ten methodological approaches with a diverse level of consistency in practical applications (Table 2). The SAM and Forum for the Future's (FFF) sustainability accounting methodologies are two of the most consistent and complete FCA approaches that have been put into practice to date which gives a strong argument for applying them in the automotive industry.

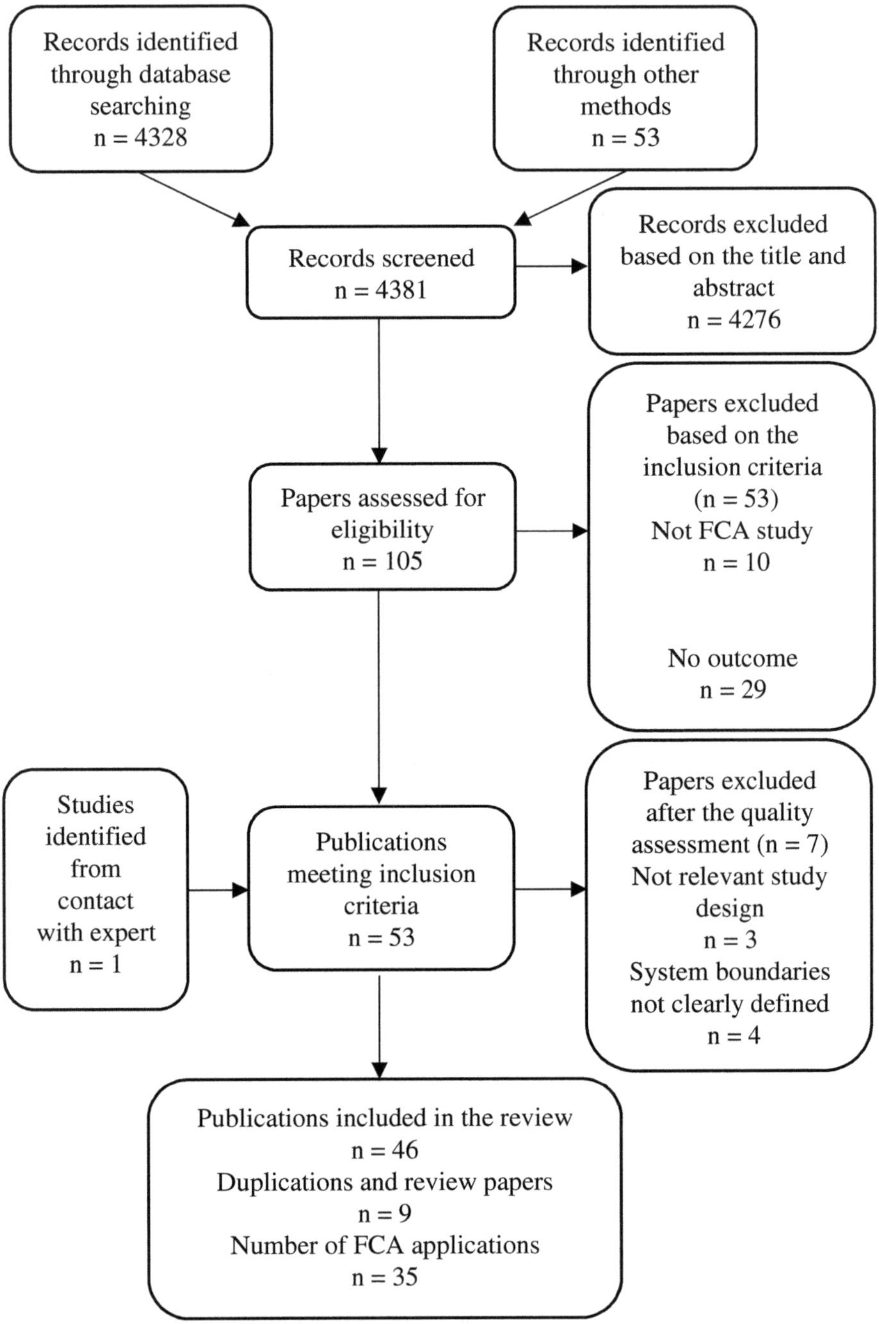

Fig. 2 Flow chart of FCA study selection process

Table 2 Methodological streams of FCA

Methodological stream/applications	Type of assessment	Cost focus	Scope (boundaries)	Type of information
1. SAM	In	Damage cost	Wide	Internal and external
2. FFF's sustainability accounting methodology	En + In	Avoidance/ Remediation cost	Narrow	Internal and external
3. United States Environmental Protection Agency's methodology for municipal solid waste management	Ec (one En study)	Market methods	Narrow and wide	Internal (One external study)
4. ExternE	En + Sc	Multiple	wide	External
5. Monetised LCA approach	En + HH	Multiple	Wide	Internal and External
6. Sustainable value concept	In	Opportunity costs	Narrow	Internal and external
7. PUMA Environmental Profit and Loss Account	En	Multiple	Wide	Internal and external
8.Ontario Hydro	En + HH	Damage cost	Wide	Internal and external
9.Extended LCC	Ec + En	Damage cost	Wide	Internal and external
10.Centre for Waste Reduction Tech-nologies	Ec + En	Damage cost	Wide	Internal and external

En Environmental, *Ec* Economic, *Sc* Social, *In* Integrated *HH* Human health

The SAM is the outcome of cooperative work between British Petroleum (BP) and the University of Aberdeen. It has been developed in order to articulate sustainability issues in a project's evaluation and operational support decisions [2]. The SAM has been applied in many different settings in the UK and New Zealand, such as the British oil and gas industry, hydrocarbon development, landfill gas (Baxter et al. [1]), organic recycling, waste management, urban transport (Cavanagh et al. [5]) and urban developments (Xing et al. [15]).

The FFF methodology is an extension of (Gray's [9]) Sustainable Cost Calculation (SCC) concept. FFF took this concept forward by adding social costs, using restoration cost as the valuation method in addition to avoidance cost and expanding the traditional profit and loss accounts and balance sheets by recognising sustainability liabilities. FFF's methodology has been implemented in organisations such as Anglian Water, Wessex Water [10], AlcCo [3] and ChemCo [14].

Both methodologies have a number of real-life applications and accessible methodology, both measure internal and external sustainability effects and use an integrated approach in the assessment although none of the FFF's studies tackled wider economic impacts and social effects were only included in the most recent applications. However, there are three major issues that prevent FFF's methodology being applied in the automotive business: the targeted audience, system boundaries and lack of flexibility.

4.1 The Targeted Audience

The FFF's methodology advocates the extension of the traditional profit and loss accounts and balance sheet so that they recognise the social and environmental liabilities. These liabilities are measured and reported during the accounting period (usually 1 year). In traditional accounting, the profit and loss accounts, balance sheet and cash flow statements are used as communication tools between a company and external stakeholders.

Management accounting creates foundations for the whole accounting system [13]. For example, annual reports contain financial information which has previously been measured on a weekly (or monthly) basis through management accounting tools. Hence, it is important to initiate changes at the management accounting level first before moving to the entity level. Organisations need to design a robust internal performance measurement system that will provide creditable and reliable data before moving to the whole system approach and reporting data to the external audience.

The SAM has been developed to articulate sustainability issues at the project assessment level and therefore it is intended for the internal audience and decision-makers [2]. The SAM supports an organisation with forward-looking information to assess potential investments, policies and help formulate strategies and assess future risks.

4.2 System Boundaries

The philosophy of FFF's methodology is that the company should be accountable for impacts that are under the control of the organisation; therefore, they are the advocates of narrow boundaries. If this approach is applied to every member of the supply chain, then every company would be equally accountable for its own impacts [10].

A number of legal requirements force automotive organisations to adopt wide boundaries to assess their products. The authors of the SAM explain that an assessment of an activity which is blind to upstream and downstream effects cannot fully address the sustainability issues. In industries such as automotive or oil and gas, the downstream sustainability effects are the outcome of decisions made before that point in time in the product chain [2].

4.3 Flexibility

Automotive organisations need sustainability cost information for a variety of decisions which include: product mix, manufacturing process design, assessment of transport modes, product disposal (recycling) strategies, comparing performance across facilities and assessing pollution prevention projects and technologies [12]. Hence, automobile manufacturers need sustainability assessment tools that are flexible to support decisions at different levels and in different configurations. The SAM has been proven to be more ideologically open and flexible than the FFF's methodology. FFF's methodology has been mainly applied at the organisational level with a single attempt to expand the idea at the product level [14]. The SAM has found practical applications in different settings and configurations including the assessment of the whole industry and at the organisational, project and material level (Baxter et al. [1]). Furthermore, there are no restrictions on taking this concept forward and extending the boundaries of the assessment at the policy, product, process or strategy level.

5 Conclusions

The systematic review of the FCA literature extracted ten methodological approaches in this field with a diverse level of consistency in practical applications. This article proposes that the SAM is the most complete and developed FCA methodology for application in the automotive industry. The SAM measures internal and external sustainability effects which can expose new business opportunities for automotive organizations such as lower costs from reduced waste sent to landfill, water and energy consumption or carbon emissions. Monitoring externalities assists in a strategic planning and risk management by anticipating future environmental and social legislation in the automotive sector. The SAM uses the life cycle approach which creates a basis for the sustainability assessment of automobiles. Hence, managers and design engineers in the automotive business can make the best possible choices as early as at the design stage by knowing all the possible economic, social and environmental costs of the vehicle throughout its lifetime. Furthermore, the SAM's ability to combine the multi-disciplinary issues of sustainability into a single framework is an effective way of communicating trade-offs and outcomes from the assessment. Having all these elements together facilitates the decision-making process by providing a more transparent and clear picture about the sustainability performance of a vehicle, production process, specific component, material or the whole organisation.

The original SAM was developed for the oil and gas industry; therefore, applying the SAM in the automotive context will require a new set of assessment criteria to be developed. Every industry varies and generates different types of social, environmental and economic effects; the SAM should therefore be adapted to reflect sustainability effects typical of the automotive industry.

References

1. Baxter T, Bebbington J, Cutteridge D, Harvey G (2003) The sustainability assessment model (SAM): measuring sustainable development performance. Offshore Eur 2:1–5
2. Bebbington J (2007) Accounting for sustainable development performance. Butterworth-Heinemann, Oxford
3. Bent D (2006) Towards a monetised triple bottom line for an alcohol producer. Using stakeholder dialogue to negotiate a 'licence to operate' by constructing an account of social performance. Sustainability accounting and reporting. Springer, Berlin
4. Burritt RL, Hahn T, Schaltegger S (2002) Towards a comprehensive framework for environmental management accounting—links between business actors and environmental management accounting tools. Aust Account Rev 12:39–50
5. Cavanagh J, Frame BR, Fraser M, Gabe G (2007) Experiences of applying a sustainability assessment model. In: International conference on whole life urban sustainability and its assessment, SUE-MoT conference proceedings, Glasgow
6. Department for Business Innovation and Skills (2013) Driving success: a strategy for growth and sustainability in the UK automotive sector. BIS, London
7. European Environment Agency (2014) Monitoring CO_2 emissions from new passenger cars in the EU: summary of data for 2013. EEA, Copenhagen
8. Fiksel JR (2009) Design for environment: a guide to sustainable product development. McGraw-Hill, New York
9. Gray R (1992) Accounting and environmentalism: an exploration of the challenge of gently accounting for accountability, transparency and sustainability. Account Organ Soc 17:399–425
10. Howes R (2002). Environmental cost accounting: an introduction and practical guide. Chartered Institute of Management Accountants
11. Maclean HL, Lave LB (2003) Life cycle assessment of automobile/fuel options. Environ Sci Technol 37:5445–5452
12. Mayyas A, Qattawi A, Omar M, Shan D (2012) Design for sustainability in automotive industry: a comprehensive review. Renew Sustain Energy Rev 16:1845–1862
13. Schaltegger S, Burritt RL (2000) Contemporary environmental accounting: issues, concepts and practice. Greenleaf Publishing Limited, Sheffield
14. Taplin JR, Bent D, Aeron-Thomas D (2006) Developing a sustainability accounting framework to inform strategic business decisions: a case study from the chemicals industry. Bus Strategy Environ 15:347–360
15. Xing Y, Horner RMW, El-Haram MA, Bebbington J (2009) A framework model for assessing sustainability impacts of urban development. Account Forum 33:209–224
16. Zokaei K, Lovins H, Wood A, Hines P (2013) Creating a lean and green business system: techniques for improving profits and sustainability. CRC Press, Boca Raton

Drivers of Sustainability in the Automotive Industry

Wilfried M. Bunzel and Thomas Ruhnau

Abstract In many cases, the increasing need to consider sustainability and environmental responsibility in the corporate strategy of automotive companies is not the original equipment manufacturers' (OEM) idea. Moreover internal and external changes force car manufacturers to strive for progress in sustainable technologies. This research provides an overview of potential drivers of sustainability in the automotive industry focusing on the OEM's view. After a short introduction, the presented conference contribution delivers an overview of aspects that lead the OEMs to concentrate on sustainability. Based on Porter's Five Forces, the authors explain actual developments which can be considered as the origins of the latest and future progress in technology. At first, the article will explain customer expatiations and needs. Afterwards, it will take a close look at potential future and current competition in order to clarify the OEMs' situation. In addition, suppliers' impact and threads from substitute products have an effect on the manufacturers' increased focus on sustainability.

1 Introduction

Managing Innovations and keeping your company up to date is a big challenge for any manager or board of directors. When it comes to Innovations and the necessity of such, clear guidelines are missing. This is even more the case when the change is driven by many different factors as within Automobile Industry when it comes to

W.M. Bunzel (✉) · T. Ruhnau
BMW Group, Petuelring 130, 80809 Munich, Germany
e-mail: wilfried.bunzel@bmw.de

T. Ruhnau
e-mail: thomas.ruhnau@bmw.com

W.M. Bunzel
Institut für Technik und Design Ingolstadt, Nördliche Ringstraße 19,
85057 Ingolstadt, Germany

I. Denbratt et al. (eds.), *Sustainable Automotive Technologies 2014*,
Lecture Notes in Mobility, DOI 10.1007/978-3-319-17999-5_13

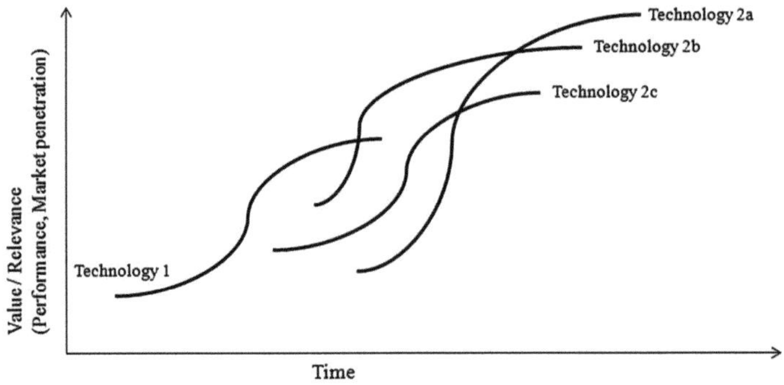

Fig. 1 S curve with several future possibilities (based on [2, 5])

sustainability. The S curve is a model which visualizes the dilemma: old technologies are well established, in many small steps improved and meet market demands. New ones are just entering the market, in the beginning with disadvantages (e.g. costs) and persuade the customer slowly before they are penetrating and improving themselves [1, 2].

Several samples in the past have shown the challenge of this step. For Example Kodak was in perfect position within photo industry and also did some research towards digital photography, but decided to stick with conventional highly developed technology [3]. In the case of Beta video tapes the case was different. A superior technology was developed and even better than competition (VHS), still market power went the other direction [4]. We see that many different factors influence the development of an industry. Therefore in many cases there exist different paths representing different technologies [5]. One has to state that no matter how much you know about the situation right now future development (time frame, steepness or even which path) cannot be predicted [6] (Fig. 1).

Once realized that we won't be able to tell which will be the right path to follow at what point in time with guarantee, we have to find a way to analyze the situation on which we can base our decision which approaches are best for our company. Porter gives a map of aspects to be taken into consideration which can guide the way [7]. Alongside these forces (customers, competition/new entries, suppliers, Substitute Products/Services) this article provides a closer look at the automotive industry with regards to sustainability from OEM's point of view.

2 Customers

Driven by marketing communication, former experiences, recommendations or personal needs, customers connect a certain expectation with the cars they buy. At this point, expectations which the OEMs need to meet are clustered in three categories which need to be explained separately:

- "Base" requirements
- "To Be" requirements
- "Differentiation" requirements

This research defines all technologies, product components and added services that provide the minimum legal level as "Base" requirements for a vehicle. Within the field of sustainability, the minimum legal level can include certain level of emissions (such as CO_2, NO_x) or safety and security requirements. Each and every customer needs the OEM to meet these base requirements because otherwise there would be no way for the manufacturer to offer the car to the market. National or international authorities would block the approval for the vehicle registration. Two famous examples are to be named: European EU 5/EU 6 emission standards according to the EU (REGULATION (EC) No 715/2007) and LEV emission standards according to the California Low-Emission Vehicle Regulations (2013).

"To Be" requirements represent the customers' expectations in terms of common achievements within the field of competition or vehicle segments. These requirements could also be called quasi-minimum requirements. They lead the manufacturers to fulfill certain levels of sustainability within production or vehicle attributes without opening a field of differentiation. Customers expect specifications and features to be accomplished, e.g. fuel consumption on a comparable level, safety features and certifications as Euro NCAP 5 star rating [8]. Depending on the segment and price class, OEMs determine adequate "To Be" requirements and handle them as legislation. The car development processes are planned accordingly.

In addition to base requirements and quasi-minimum requirements, an OEM has the opportunity to define a field of differentiation versus the competitors to meet or over-achieve customer expectations. In terms of sustainability, the OEMs can focus on monetary benefits for the customer to gain advantage over competitors and the manufacturer can strive for an improvement of brand image. At a generic level, monetary benefits lower the total cost of ownership (TCO). Lower fuel consumption leads to savings at the gas station, lower emissions can provide lower taxes on a monthly/yearly basis as well as at the time of purchase. Fuel efficiency can even have an influence on personal income taxes. In addition to taxes, low emission or electric vehicles are sometimes subsidized with lump sum payments by the government to encourage customers to decide for an environmental friendly car. Legislation and taxation differ from country to country and thresholds to achieve the next level of support are not internationally harmonized, either. In consequence, OEMs need to take into account the specific international market penetration and technological skills compared to the specification of competitor cars in order to define the best differentiation requirements to meet. Next to monetary benefits as differentiation requirements, there is also the chance to develop achievements in a qualitative regard. Usually, the customer does not profit from improved sustainability technologies in production, material or logistics. Nevertheless, the customer's expectation towards automobile brands to understand sustainability in a holistic way opens a field for the OEMs to differ from each other. Depending on the OEM's specific international market shares, brand and image policy, technological advantage, the differentiation requirements are set to get the maximum effect on the fulfillment of customer expectations at a minimum effort. As a result, customer expectations, which are partly influenced by legislation and taxation, are a major driving force for advanced achievements in automotive sustainability.

3 Competition

First, this research takes a look on industry-internal competition; afterwards potential new competitors are discussed.

Both, premium and base segment OEMs provide similar goods and services, they cannot increase or decrease their capacities or change the industrial sector in a short or mid-term way. In consequence, this leads the automotive industry to be characterized by an intensive competition which all OEMs have to consider in their strategies [9]. As manufacturers try to maintain or enlarge their market share, they strive for improving their products in order to achieve an advantage versus their competitors. Next to classical automotive related differentiators, such as horsepower, trunk size, riding comfort etc., more and more customers understand sustainability as a value. Especially technologies, which lead to a monetary benefit, are attractive to the consumer (see above). But not only financial advantages are important, also environmental friendly production, materials or recycling technologies find their way into the customer's mind set. By focusing the development of cars on sustainability, OEMs can find a new field of differentiation versus the competition including potential price advantages [10]. As soon as the competition achieves a new level of sustainability (e.g. in CO_2 emission) and communicates the value to the customer, other OEMs need to follow in order not to lose attractiveness. The same is valid for achievements in technology. In consequence, it can be concluded that achievements in sustainability within the competitive environment is a driver of further enhancements in this field of technology. In this context, it can easily be observed that the driving forces Customer and Competition are closely connected to each other as interdependencies occur over time.

Within the automotive industry, the field of competitors is well known and the OEMs use standardized processes of competitor analysis [11]. Of course, proceedings in sustainable technologies within the common field might also be a threat for the current market position. But major impacts on the industry can be caused by new competitors. As described at the beginning of this article, new technologies replace established ones and often these new technologies get the necessary maturity by the development of new competitors. One famous example is given by the proceedings in e-mobility and the corresponding success of Tesla Motors. Founded in the early 2000s, Tesla is a very young brand but they boosted registration of their electric cars up to over 20,000 units in 2013. This example shows that new technologies developed by new players can influence the market considerably which forces established OEMs to look beyond their own nose not to miss a trend or technology. This is also the reason why some OEMs cooperate with the estimated origin of further advances in sustainable drive train technology—battery companies [12].

4 Suppliers

A share of about 50 % of research of the automotive industry is done by suppliers [13]. This emphasizes the big role within development of new technologies suppliers play also when it comes to sustainability. Besides if an OEM wants to implement a holistic sustainable approach, suppliers must not be forgotten. So they start to push them into ISO certifications and to keep their company policies or request standards already within purchasing process [14]. But to come to the next level also second-tier suppliers and beyond must be included [15]. Even other suppliers like service providers and marketing agencies as well as all processes can and should be taken into consideration as shown by Weber et al. [16].

So suppliers have to be ready for a change. They basically face the same situation: What technology will be relevant at which point in time? If they act early they have the chance to set standards and to get new customers. On the other hand there is the risk of a wrong investment. A further challenge for the whole industry is a sudden increase in demand for a technology when there is not enough supply overall as it can be seen in terms of LED these days [17].

5 Substitute Products/Services

As pointed out by Porter one must not only consider competition and new competitors, but also all relevant substitute products and services. We see a trend in the automotive industry these days, that this section gets more and more relevant [18] as even OEMs such as BMW and Mercedes start to offer car sharing [19, 20]. Even expectation is higher on OEMs' offer this move allows getting completely new customers [21]. Car sharing is attractive to people who live in urban areas where they can usually use public transport systems and do not find parking lots easily. So they have a relatively cheap possibility to have a car available for example to transport shopped goods or to have a comfortable private ride [22]. Still car sharing is not available everywhere and smaller cities have a lower range of offer [23]. Car pooling is maybe not to be considered as in most cases all participants need to have a car to share driving duties. But agencies for arranging rides need to be considered. In these cases transportation from one city to another is offered as service by private persons with their own car [24]. On the other hand we have public transport systems like bus (in town, region or even nationwide), plane, railway, suburban railway and subway. Besides car rental and taxis could be considered as well. The difference

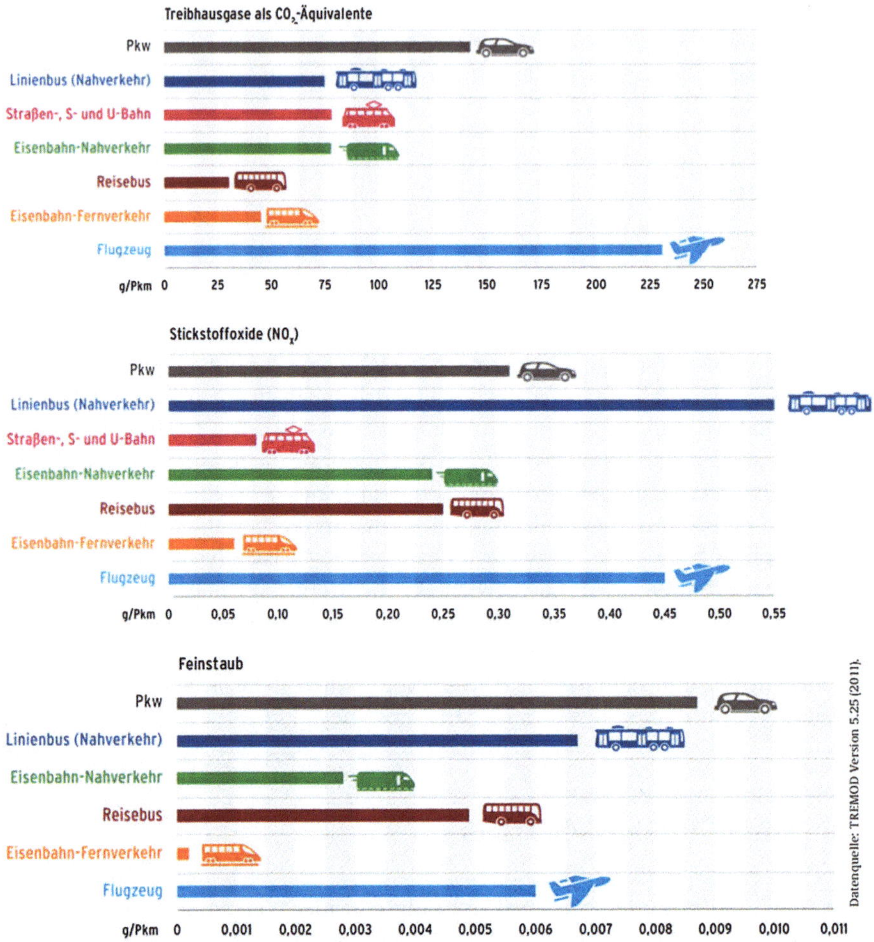

Fig. 2 Overview about environmental impact [18]

between car sharing and rental can be focused on 24/7 pick-up, charging for using time, fuel included within price and different pick-up/give back positions [25, 26].

As shown there are several different options to choose from if one would like to avoid having or using a personal car. But what is the most environmental friendly one today. Figure 2 gives an overview showing CO_2-effect, nitrogen oxides and particulate matter per person kilometer in gram for car, bus (in town), subway, train short range, bus between towns, train long range and plane [18].

VCÖ shows and stresses that not only the greenhouse gas that is emitted while moving is to be considered but also the one for providing fuel, production and proper disposal of transportation means [27] (Fig. 3).

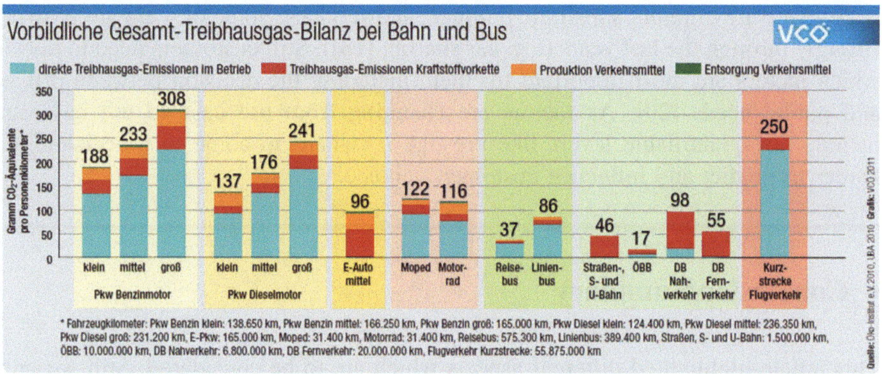

Fig. 3 Environmental impacts along the whole lifetime of transportation means [27]

But sustainability or environmental friendliness is not the only factor that will be analyzed by a consumer when it comes to the decision for any of the possibilities or the decision not to buy a (further) car at all. Time and costs are maybe the major other drivers. Figure 4 shows how these vary through different offers and for different use cases.

Strecke innerstädtisch **Berlin: Schlesisches Tor (Kreuzberg) – Humboldt-Universität (Unter den Linden/Mitte)**	**Fahrrad**	**ÖPNV (Bus & Bahn)**	**Pkw (1 Person)**	**zu Fuß**
Entfernung	4,0 km	5,9 km	6,5 km	4,0 km
Kosten	0,36 Euro	2,10 Euro	3,64 Euro	0,00 Euro
Zeit	14 min	26 min	23 min	49 min
CO_2-Ausstoß	0,00 kg	0,42 kg	1,12 kg	0,00 g

Strecke Berlin–Frankfurt (jeweils Stadtzentrum)	**Bahn**	**Pkw**	**Flug**	**Reisebus**
Entfernung	510 km	543 km	433 km	543 km
Kosten	ab 29 Euro	305 Euro	ab 33 Euro	33 Euro
Zeit	4:19 h	5:04 h	5:10 h	9:50 h
CO_2-Ausstoß	26,0 kg	94,2 kg	81,2 kg (ohne RFI-Faktor*)	19,2 kg

Fig. 4 Distance, cost, time and CO_2-effect of bike/bus/car/walk on short distance and of train/car/plane/bus on long range [28]

So there are different substitute products and services. Some of them are growing strongly through the last years (e.g. car sharing [18]). Still customers need to decide which factors are most important for them including the decision between private and public needs [29]. As factors are changing from use case to use case and include not quantifiable levers like prestige it is hard to come to a decision and harder to predict and influence customers' choice.

6 Conclusion/Summary

This article highlighted different aspects which are to be considered. Still we only will be able to know afterwards what would have been the perfect strategy. Right now it seems a good strategy to be prepared to act in different technologies/services and start using these channels already today especially when this is possible with a profitable approach yet.

References

1. Abernathy WJ, Utterback JM (1978) Patterns of industrial innovation. Technol Rev (Cambridge) 80(7):40–47
2. Rafinejad D (2007) Innovation, product development and commercialization. J Ross Publishing, Fort Lauderdale
3. Nunes P, Breene T (2011) Jumping the S-curve. Harvard Business Review Press, Accenture, Boston
4. Timmons MB, Weiss RL, Loucks DP, Callister JR, Timmons JE (2014) The entrepreneurial engineer. Cambridge University Press, New York
5. Foster RN (2006) Innovation: die technologische Offensive, McKinseyClassics 6. Redline Wirtschaft, Heidelberg
6. Floyd C (1997) Managing technology for corporate success. Gower, Hampshire
7. Porter ME (1996) On competition. Harvard Business Review, Boston
8. EuroNCAP (2014). http://www.euroncap.com/test/ratings.aspx. Accessed 30 April 2014
9. Becker H (2007) Auf Crashkurs. Springer, Heidelberg
10. Ruhnau T, Bunzel WM (2014) Sustainability in automotive pricing. In: Wellnitz J, Subic A, Trufin R (eds) Sustainability automotive technologies 2013. Springer International Publishing, Switzerland
11. Diez W (2006) Automobilmarketing - Navigationssystem für neue Absatzstrategien, mi-Fachverlag, Landsberg am Lech
12. Bundesministerium für Bildung und Forschung (BMBF) (2010) Elektromobil in die Zukunft - Batterieforschung als Schlüssel, Bonn Berlin
13. Minhas SUH, Lehmann C, Berger U (2011) Concept and development of intelligent production control to enable versatile production in the automotive factories of the future. In: Hesselbach J, Herrmann C (eds) Globalized solutions for sustainability in manufacturing. Springer, Berlin
14. Stead WE, Stead JG (2004) Sustainable strategic management. M. E. Sharpe, New York

15. Lee S-Y, Cheong I-M (2012) Sustainable supply chain initiatives in the Korean automotive industry. In: Madu CN, Kuei C-H (eds) Handbook of sustainability management. World Scientific Publishing, Singapore
16. Weber J, Bowler M, Kurfess T (2011) A holistic approach to sustainability evolutions in the automotive industry. In: Hung S, Subic A, Wellnitz J (eds) Sustainable automotive technologies 2011. Springer, Berlin
17. Chips and More (2014). http://www.chipsandmore.net/kundenportal/index.php?option=com_content&task=view&id=102&Itemid=180&lang=es. Accessed 21 April 2014
18. Umweltbundesamt (2012) Daten zum Verkehr, Ausgabe 2012, Dessau
19. Drive Now (2014). https://de.drive-now.com/. Accessed 19 April 2014
20. Car2Go (2014). https://www.car2go.com/de/berlin/. Accessed 19 April 2014
21. Diehlmann J, Häcker J (2013) Automotive management. Oldenbourg Wissenschaftsverlag, München
22. Transportation Research Board (2005) TCRP report 108, car-sharing: where and how it succeeds. Transportation Research Board, Washington
23. Carsharing-Experten.de (2014). http://www.carsharing-experten.de/. Accessed 19 April 2014
24. Mitfahrzentrale (2014). http://www.mitfahrzentrale.de/. Accessed 19 April 2014
25. Groß S (2011) Tourismus und Verkehr: Grundlagen, Marktanalyse und Strategien von Verkehrsunternehmen. Oldenbourg Wissenschaftsverlag, München
26. Auto Zeitung (2014). http://www.autozeitung.de/finanzen/was-ist-carsharing-begriff-erklaerung-unterschied-mietwagen. Accessed 19 April 2014
27. VCÖ (2011) Factsheet, VCÖ: Gesamtbilanz zeigt die wahren Umweltschäden durch den Verkehr. VCÖ, Wien
28. VCD (2014). http://www.vcd.org/verkehrsmittel-vergleich.html. Accessed 19 April 2014
29. OECD (2002) Towards sustainable household consumption? Trends and policies in OECD Countries. OECD Publications Service, Paris

Part V
Material-Light Weight Design

Aspects of Damage Tolerance and Fatigue of CFRP Structural Components

U. Burger and L. Rochat

Abstract Lightweight structures are one key issue for all future mobility concepts. Carbon fibre reinforced plastics (CFRP) play an important role in these disciplines due to their outstanding mechanical performance regarding to their weight. Therefore, CRFP structures have been widely used since decades in aerospace industry resulting in improvements in payload, fuel consumption and range. The Airbus A350, Boeing B787 in civil airplane industry as well as military products like the NH90 transport helicopter are examples of this development towards "all composite"-aircrafts with nearly all structural parts made out of composite materials. The increasing importance of electro mobility in the automotive sector leads to the necessity of new structural design approaches apart from commonly used metallic designs. The objectives are to reach adequate ranges by compensating for the additional weight due to electric components of hybrid or purely electric driven vehicles by improved lightweight structural design. The technological peak of this development are the i-models of BMW using CFRP for body-in-white-structure for the first time in commercial automobiles. A main difference of CFRP-structures towards metallic ones is the behavior regarding damage tolerance and fatigue. For helicopter composite structures this issue is newly defined in §573 of the relevant certification specifications (CS for EU, FAR for US) valid since December 2012. This paper for the first time summarizes the necessary efforts and possible methods to show compliance for this specific topic in the aerospace industry. As there is no similar specification or rule available now, this may serve as an example in other fields like the automotive industry. This paper discusses approaches used in aerospace industry to design robust and lightweight CFRP structural components, which can also be transferred to automobile structures. This includes definitions and standard practices defining the damage, generation of design allowables and necessary testing effort.

U. Burger (✉)
Technische Hochschule Ingolstadt, 85049 Ingolstadt, Germany
e-mail: uli.burger@thi.de

L. Rochat
Institut für Technik und Design, 85057 Ingolstadt, Germany
e-mail: ludovic.rochat@itd-in.de

I. Denbratt et al. (eds.), *Sustainable Automotive Technologies 2014*,
Lecture Notes in Mobility, DOI 10.1007/978-3-319-17999-5_14

1 Introduction

Over the last decades, composite materials have increasingly been used in the aerospace industry: the high specific strength and stiffness, the resistance to corrosion, reduction of number of parts, crashworthiness, fatigue resistance and tailorable properties make polymeric composites very attractive [1].

The Windecker Eagle made of GFRP was the first all-composite aircraft to comply with the FAA regulations and to be certified in the US in 1969. Since then and with the introduction of advanced composites in the 1980s such as the CFRP, these materials are no more only used for secondary loading structures such as the rudder, spoilers, airbrakes, elevators and flaps of the A320. Applications nowdays include the wing box and fuselage of the new generation of long-range transport aircraft including the B878 and A350 whose percentage of their structural weights attributed to composites are around 50 % (see Fig. 1).

Composite materials are also used in other industries such as the energy (wind turbines), marine, rail transport as well as in the automobile industry. Looking at the car history, the first car made of fiberglass might be the Chevrolet Corvette introduced in 1953. Nowadays, the average proportion of plastics in the automobile structure is around 16 % and could reach 18 % by 2020 [3]. Monocoque passenger cell, seat shell, roof module, pedals, airbag housing, engine air intake manifold, air filter system housing,, bumpers, front end, engine mounts, spare wheel recess are common parts made of composites [4].

Weight reduction is the principal reason of using lightweight materials. Since 2010 the average total weight of the vehicle seems to reduce [3]. This new trend might come from the new regulation in the U.S. and E.U. regarding respectively the

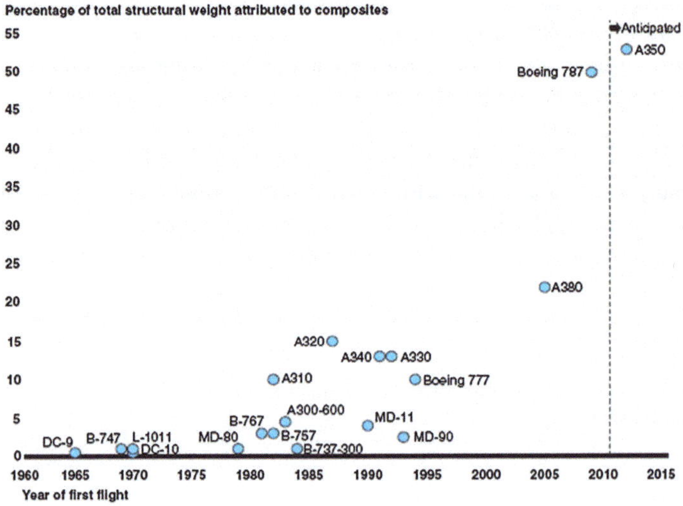

Fig. 1 Proportion and evolution of composites in commercial airplanes [2]

fuel consumption efficiency (proposed standard from 34.1 mpg in 2016 to 49.7 mpg in 2025 for combined passenger car according to [5]) and the CO_2 emission (95 g CO_2/km target by 2020 [6]). A weight reduction of 100 kg could lead to a drop of 6 g CO_2/km [7]. Moreover, the interest for lightweight structures is amplified by the new electric power trains which results in heavier cars (the electric power systems account for 20 % of the total weight of the vehicle instead of 12 % for the conventional engine due to heavy batteries) [3]. Up to now, plastic composites are mainly used in for the race cars (Formula 1) and low-volume sport and luxury cars (such as the Lamborghini Aventador, MacLaren MPC and VW XL1).

It is hardly predictable to know whether or not, and in which proportion, the automobile industry is going to implement composite parts in the new coming designs. Many challenges remain to be addressed to use composites for high volume production, such as cycle time (less than 1 min per cycle is required), recycling (thermoset plastics are difficult to reclaim), availability (limited numbers of qualified suppliers and volatile market) and cost (10 times more expensive than conventional materials) [8]. However, the interest for composites from the automotive industry has never been so high. Many partnership between these two industries, whose aims are to develop either new manufacturing processes (High Pressure RTM) or composite materials requiring lower cycle time (such as the thermoplastic matrix) and less expensive to manufacture, have recently been created. BMW Group and the carbon fibre manufacturer SGL Group (develop high pressure RTM and fast-cure epoxies), General Motors and Teijin (develop thermoplastic matrices), Daimler AG and Toray Industries (using RTM process) as well as Ford Motor Co. and Dow (develop cost-effective carbon fiber composites).

The automakers BMW is one of the biggest users of CFRP with its all-electric I3, the first time CFRP is used for most of the internal structure of a mass production passenger car (see Fig. 2).

Even if the automobile and aerospace industries deal with different technological challenges, one might wonder how and to what extend the experience in composites from the aerospace could provide solutions to automobile applications. As shown by the new airworthiness regulations, the degree of maturity of composite keeps increasing: less and less used as "black metal", structural optimization methods are leading to even more lightweight structures.

Fig. 2 Lightweight passenger cell of BMW I3 made of carbon (*Source* BMW website)

2 Regulations in the Aerospace Industry

The regulations CS for the EU and FAR for the US provided respectively by the airworthiness authorities EASA and FAA, usually harmonised, and regularly updated and refined. For an aircraft to be certified in the relevant countries, its design has to show compliance with these regulations. The four documents CS (FAR)23-25-27-29 deal with regulations for respectively small aeroplanes, large aeroplanes, small rotorcraft and large rotorcraft.

2.1 Definitions

The following definitions, generally common to all the CS and FAR documents, are given below to clarify the airworthiness regulations in terms of fatigue evaluation discussed in the next paragraph [9–11]:

- Damage tolerance: The attribute of the structure that permits it to retain its required residual strength for a period of use after the structure has sustained a given level of fatigue, corrosion, or accidental or discrete source damage.
- Fail-safe: The capability of structure remaining after a partial failure to withstand design limit loads without catastrophic failure within an inspection period.
- Safe-life: The capability of as-manufactured structure as shown by tests, or analysis based on tests, not to initiate fatigue cracks during the service life of the rotorcraft or before an established replacement time.
- Fatigue safe-life: Structure must be shown to be able to withstand the repeated loads of variable magnitude expected during its service life without detectable cracks.
- Design limit load: The maximum loads to be expected in service.
- Ultimate load: Limit load multiplied by prescribed factors of safety.
- Residual strength: The strength retained for some period of unrepaired use after a failure or partial failure due to fatigue or accidental or discrete source of damage.
- Principal structural element (PSE): A structural element that contributes significantly to the carrying of flight or ground loads and whose failure (such as fatigue cracking) can lead to catastrophic failure.
- Catastrophic failure: Any structural failure, which results in death, severe injury, or loss of the aircraft.
- Accidental damage: Discrete damage that may occur in service use or in manufacturing due to impacts or collisions, such as dents, scratches, gouges, abrasions, disbonds, splintering and delaminations.

Principal structural elements (PSE) include all structural elements susceptible to fatigue cracking which can lead to catastrophic failure such as:

- Aeroplanes: wind and empennage (control surfaces, spars, skin-stringer combination, primary fittings and spliced), fuselage (circumferential frames, door frames, window frames, pressure bulkhead), landing gear and engine mounts

- Rotorcraft: rotor blades including the attachment fittings, rotors head (hub, dampers, hinges), rotor supporting surface, landing gear, fuselage and auxiliary lifting surfaces.

2.2 Damage Tolerance and Fatigue Evaluation

Damage tolerance and fatigue evaluation of composite structures are addressed in §573(a) for CS(FAR)-23, in §573 for CS(FAR)-27 and 29 and in §571 for CS (FAR)-25.

The regulations are accompanied by Acceptable Means of Compliance (AMC for CS) and Advisory Circulars (AC for FAR). These documents provide example of acceptable means of showing compliance. Even if they are not mandatory, the guidance reflect the present good design practices and are generally followed by the aerospace manufacturers in order to facilitate the certification. The means of showing compliance AC No 20-107B and AMC 20-29 deal specifically with composite aircraft structure [12].

The evaluation of the composites structure must show that no catastrophic failure will occur during the operational life of the aircraft and has to deal with the static and repeated loads, adverse environmental effects, manufacturing defects and accidental damage. The so-called "damage tolerance fail-safe" and "fatigue safe-life" evaluations (defined above) are the two possible design approaches. The latter is considered only if the former is "impractical within the limits of geometry, inspectability, and good design practise" [13]. The damage tolerance evaluation should include the PSE.

The damage tolerance evaluation procedure is given below:

- Identification of the parts being evaluated
- Determination of the critical load cases expected in service including the fatigue (repeated) load spectrum that contribute to damage growth
- Damage threat assessment: determination of the types of damage that might occur during manufacturing, maintenance or operation. Depending on the damage severity and considering the tools available for inspection
- Determination of the residual strength of the structure: one should demonstrate that the residual strength of the structure will be above the strength required to withstand the design loads. The sensitivity of the structure to damage growth has to be controlled (crack stoppers and multiple load paths) to ensure adequate residual strength. It has to be ensured that the residual strength doesn't remain during an excessive period of time below the strength required to retain ultimate load (see Fig. 3).
- Replacement times and inspection: frequency, extent and methods defined the maintenance plan whose goal is to restore ultimate load capability of the

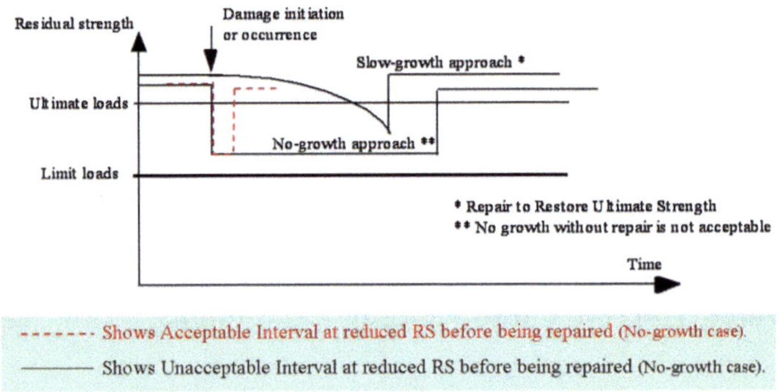

Fig. 3 Schematic diagram of residual strength accounting for damage occurrence and repair [12]

structure. The inspection interval depends on the damage detectability threshold, depending itself on the inspection methods used. Inspection intervals should be set between the time the damage is initially detectable and the time the damage severity reaches the limit corresponding to critical residual strength capability (see Fig. 4).

- Effect of damage on the stiffness and dynamic effects should be considered since strength is not always the driving criterion to size structures such as the wings (stiffness-based criterion).

If the damage tolerance fail-safe evaluation is impractical then a fatigue safe-life evaluation is then performed. Fatigue substantiation should account for the effects of the environment and for the flaws and accidental impacts that might occur during manufacture, maintenance and operation.

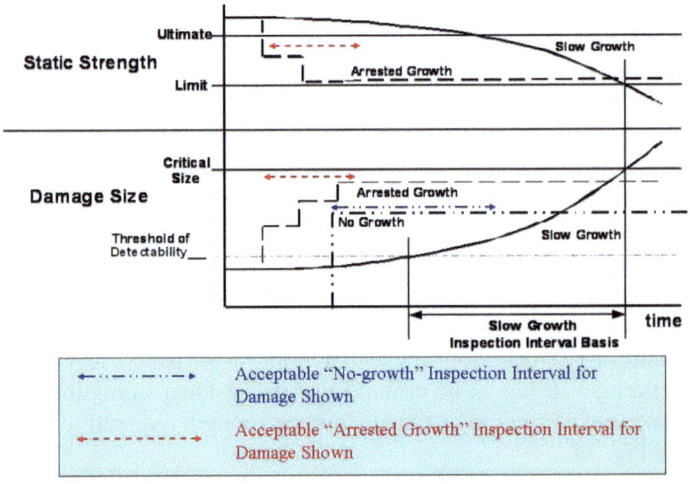

Fig. 4 Residual strength and damage size relation, acceptable inspection intervals [12]

3 General Approach

3.1 Damage Tolerance and Fatigue for CFRP Versus Metals

The procedure presented above in terms of damage tolerance and fatigue evaluation is similar for metallic and CFRP structures (critical loading, threat assessment, determination of residual strength and inspection programme). The difference remains in the way the procedure is implemented since the impact responses and fatigue characteristics (damage initiation/evolution) have nothing in common: whereas their assessment for metallic parts is rather straightforward and relies on empirical equations from tests and experience, no database is available for composites due to the variety and complexity of their failure modes (see Fig. 5).

Moreover, CFRP structures are very sensitive to out-of-plane loads. Low velocity impacts (such as dropped tools) are of great concerns since they are likely to be undetected (barely visible impact causing internal damage such as delamination) while decreasing significantly the load-carrying capacity of the structure [15–17].

3.2 Threat Assessment

Starting point for the damage tolerance methodology must be a proper assessment of realistic impact energy levels. A part-by-part analysis is recommended as threats may easily differ regarding to the design properties, location and functions of the PSE. Data may be derived out of in-house experience, customer support data or from literature. Regarding H/C-structures (sandwich), an overall assessment is given in [18] and shown in Fig. 6.

In this diagram different identified impact threats are given with their corresponding energy level and their probability. For instance, 99 % of the terrain object impacts occurred at max. 50 J. This has then to be matched with the single PSE whether it is feasible or not. Deeper insight regarding single models or components are derived by in-house-investigation and customer support data.

	Metal	Composite (CFRP)
Crack initiation	Stress concentration, voids	Stress concentration, voids, micro-cracks, fiber debonding, fiber discontinuities, misalignment
Crack propagation	Single crack growing	Propagation of multicraks through matrix and fibre/matrix interface
Final failure	Clean fracture surface	Combination of matrix cracks, debonding, delamination (sublaminate buckling) and fiber breakage

Fig. 5 Metal fatigue compared with composite fatigue [14]

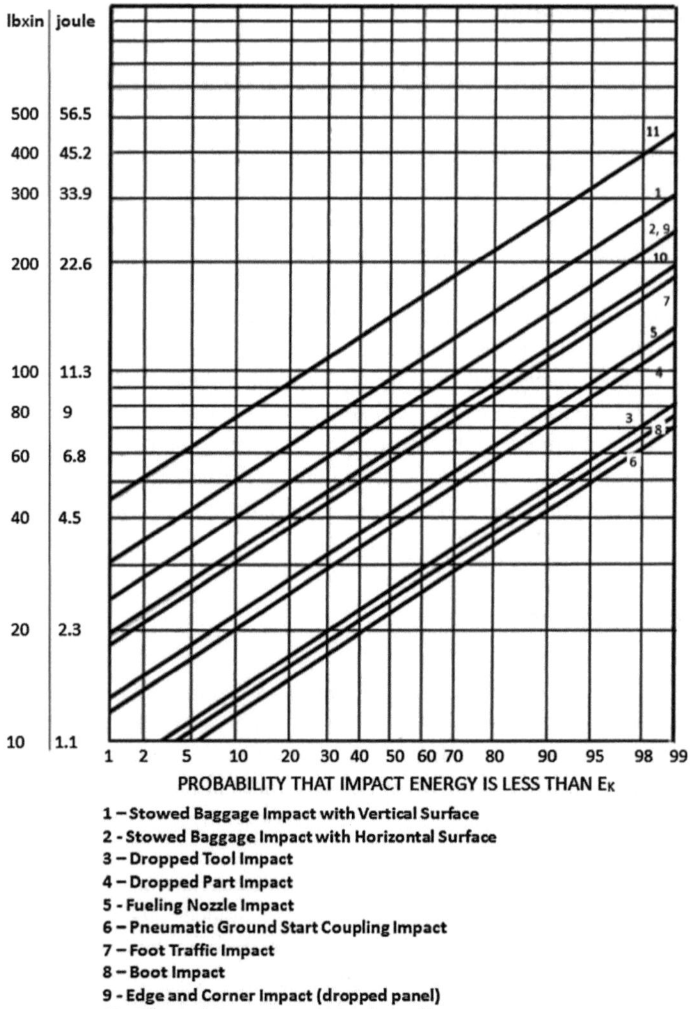

Fig. 6 Possible impact energy level for airframe structure [18]

3.3 Detectability Thresholds

As stated before damage detection for composite structure points out to be a real challenge because of the possibility of large invisible damage inside the structure which cannot be seen from outside. For practical reasons only means of visible inspection should be considered for defining the detectability of damages. Different methods and definitions are given in [19]. A possible mean for defining detectability of damage is the dent depth after an impact event. An investigation by

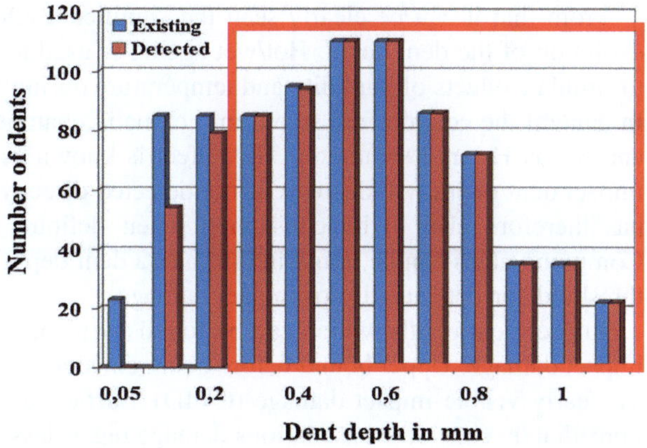

Fig. 7 Dent depth detectability by detailed visual inspection [21]

Airbus shows that dent depths of 0.3 mm and higher are detected by 100 % (Fig. 7) and can serve as a first value for detectability of damage. The investigation was carried out with blue collar workers and by so called "detailed visual inspection". This means close proximity to the structure (max. arm length), possible cleaning of surface, sufficient and if needed additional light and simple additional means like magnifying lenses.

This threshold is known as barely visible impact damage (BVID). This value is also given in [20] where testing procedure for composite coupons is given. The influence of boundary conditions such as time, loading and environmental effects has been studied by another investigation. The results are shown in Fig. 8.

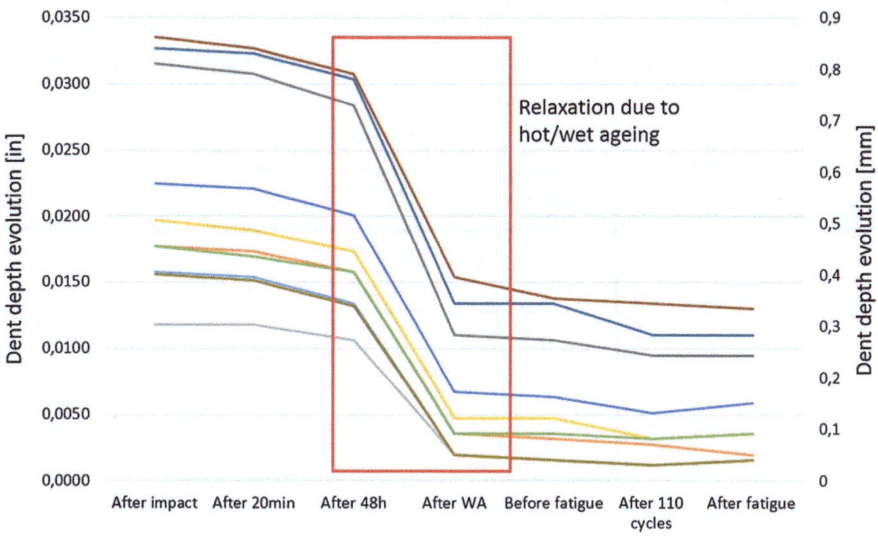

Fig. 8 Relaxation effects on dent depth [21]

From that it can be clearly seen the significant effect by hot/wet ageing to the evolution of the dent depth. Hot/wet ageing is used as a laboratory artificial means to simulate effects of humidity and temperature during life time for composite parts. In general the coupons are stored in a climatic chamber for 1000–1500 h at 70 °C and 85 %r.H. until saturation. This effect is known as relaxation meaning that the impact dent depth will decrease if not detected directly after the impact event. This has therefore also to be considered when defining the detectability threshold. Common values can be found in [19] and a dent depth of 1 mm commonly used as threshold for detectability of impact damage.

Impact dent depths which can be found by inspection are considered as visible impact damage. Upper bound before being considered as obvious damage is known as clearly visible impact damage (CVID). Surface damage like fibre breakage are immediately considered as obvious damage regardless of the dent depth and require more detailed inspection and repair.

The corresponding energy level necessary to obtain the dent depth is strongly dependent to the layup and material used and must therefore be defined individually for every PSE.

3.4 Zoning Diagram

Putting together the information of threat assessment and detectability of damage leads to a zoning diagram [19] in general which can then be applied to every PSE individually (Fig. 9).

Damage caused by possible threats with a dent depth lower than the BVID threshold is considered as undetectable. Therefore ultimate load capability is necessary together with a no-growth of damage during the whole life-time of the structure. This zone is called Zone 1. The region of damages with dent depths

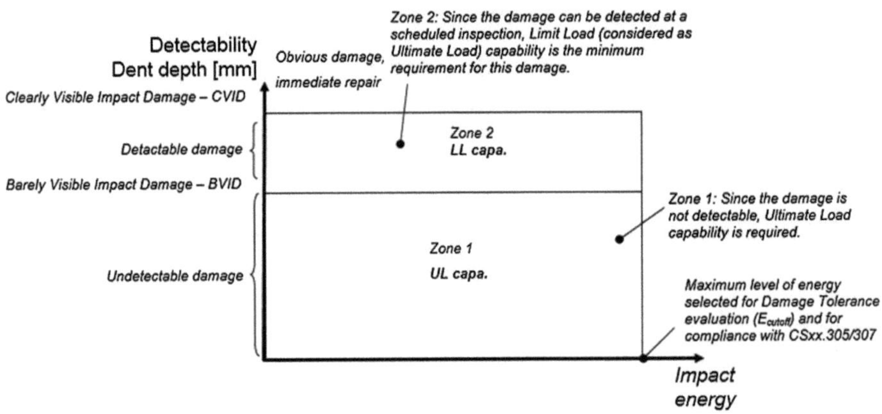

Fig. 9 General zoning diagram for PSE according to [19]

between BVID- and CVID-level is called Zone 2 and sets to be the damage tolerant zone. Within this zone the structure must have limit load capability and no-growth of damage during one inspection interval is mandatory. During inspection damage will be found and ultimate load capability will be restored by repair or change of part. Damages outside this region is either obvious and will require immediate repair or beyond the possible impact energies derived from the threat assessment.

4 Design Allowables

Allowables for design process are important for reducing testing effort and confidence in selected design. Regarding damage tolerance of composite structures general strain criteria is widely used in aerospace industry. The necessary strain level is derived by compression testing of coupons including damage, so called compression after impact (CAI), as this is the most critical loading for composite structures used in aerospace structures. Coupons used must be representative of the PSE structure in shape and layup. General allowables for preliminary design may be derived by generic coupons. Tension and shear testing of damaged coupons here in general give higher allowable strain values. This includes also testing considering environmental effects.

No growth of damage muss be shown when impacted coupons are dynamically loaded in compression with constant amplitude of a certain percentage of σ_{CAI} (static residual strength). Periodically the damaged area is measured by NDT such as US scan to control damage size, see Fig. 10.

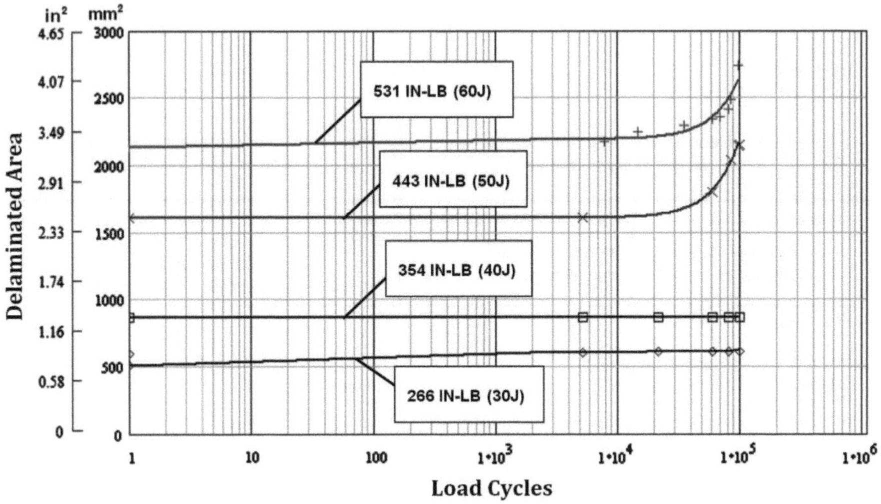

Fig. 10 Damage progression of CFRP coupons [21]

Coupons not destroyed by this test sequence are again tested for their compression strength now called compression after impact and fatigue (CAIF). For the static test environmental effects has to be considered as well. Out of CAI and CAIF results the design allowable is then derived.

5 Full Scale/Component Testing

General test procedure for full scale resp. component testing of new composite structure is shown in Fig. 11.

The structure used for the test must be representative, thus including possible manufacturing defects, BVID, repairs, etc. At the beginning of the procedure static limit load resp. a percentage of it is applied. This also serves as a check for FEA. After that one design service goal will be tested by typical spectrum loading including additional factors for e.g. temperature effects, statistical effects, etc. After that static ultimate load test takes place. Before next test sequence CVID are included in the structure. Then one inspection interval is tested with typical spectrum loading, again with additional factors. After that another static test is performed showing at least limit load capability.

If the designed structure is similar to already proven and known design and sufficient experience can be shown there is the possibility to perform an alleviated test program [11]. Similarity includes loading and load levels, material, lay-up, manufacturing processes and design features. Within this test procedure the first part of the general test procedure can be skipped and only the second part starting with static ultimate load test has to be performed, see Fig. 12.

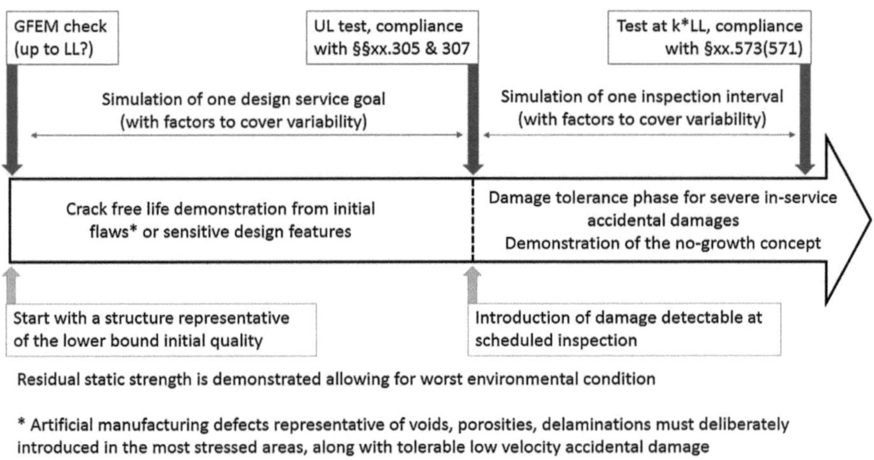

Fig. 11 Full scale/component standard test procedure [21]

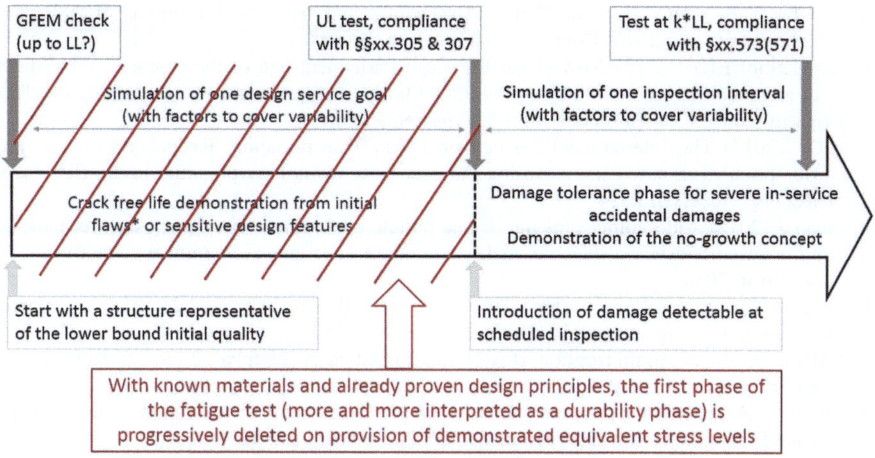

Fig. 12 Full scale/component alleviated test procedure [21]

6 Conclusion/Summary

The paper summarizes the approach towards damage tolerance and fatigue evaluation of structural components of aircraft airframes. It can be concluded in the following major steps:

- Definition of principle structural elements (PSE)
- Threat assessment to define the possible impact levels
- Definition of detectability levels by inspection
- Zoning of PSE to define necessary load capability
- Coupon testing for definition of design allowables
- Full scale/component testing

References

1. Saligheh O et al (2012) The study of response of high performance fiber-reinforced composites to impact loading. In: Tamin MN (ed) Damage and fracture of composite materials and structures. Advanced structured materials, vol 17. Springer, New York, pp 129–141
2. GAO (2011) United States Government Accountability Office, report to congressional requesters, aviation safety. Status of FAA's actions to oversee the safety of composite airplanes, 2011, United States
3. Weill D, Besland L et al (2012) Plastics. The future for automakers and chemical companies. A.T. Kearney
4. Vink D (2012) Black carbon fiber appearing in car bodies. In: European plastics news. http://www.europeanplasticsnews.com/subscriber/featured2.html?cat=1&featuredid=1752. Accessed 2014

5. Yacobussi B, Canis B et al (2012) Automobile and truck fuel economy (CAFE) and greenhouse gas standards. Congressional research service
6. Regulation (EU) No 333/2014 of the European Parliament and of the Council of 11 March 2014 amending Regulation (EC) No 443/2009 to define the modalities for reaching the 2020 target to reduce CO_2 emissions from new passenger cars
7. ICCT (2013) The International Council on Clean Transportation. Reducing CO_2 and fuel consumption from new cars: assessing the near-term technology potential in the EU. www.theicct.org. Accessed 2014
8. Sloan J (2012) Auto composites quest: one-minute cycle time?. In: Composites technology. http://www.compositesworld.com/articles/auto-composites-quest-one-minute-cycle-time. Accessed in 2014
9. Federal Aviation Administration (FAA) (2011) Advisory circular 25.571-1D. Damage tolerance and fatigue evaluation of structure
10. Federal Aviation Administration (FAR) (2008) Advisory circular 29-2C. Certification of transport category rotorcraft
11. European Aviation Safety Agency (EASA) (2012) Certification specifications and acceptable means of compliance for large aeroplanes CS-25, Amendment 12
12. European Aviation Safety Agency (EASA) (2010) Acceptable means of compliance AMC 20-29. Composite aircraft structure
13. European Aviation Safety Agency (EASA) (2012) Certification specifications for large rotorcraft CS-29, Amendment 3
14. Niu M (2010) Composite airframe structures, 3rd edn. Hong Kong Conmilit Press Ltd., Hong Kong
15. Chai GB, Foo CC et al (2008) A model to predict low-velocity impact response and damage in sandwich composites. Compos Sci Technol 68:1348–1356
16. Klaus M, Reimerdes HG et al (2012) Experimental and numerical investigations of residual strength after impact of sandwich panels. Int J Impact Eng 44:50–58
17. Abrate S (2001) Modeling of impact on composite structures. Compos Struct 51:129–138
18. NTIS Report. Nb. ADA087609 "Advanced structures maintenance concepts"
19. Composite Materials Handbook—CMH17, vol 3
20. ISO 18352, Carbon-fibre-reinforced plastics—determination of compression-after-impact properties at a specified impact-energy level
21. Engleder A, Francescatti D, Burger U (2014) A damage tolerance and fatigue evaluation approach for composite rotorcraft airframe structures. In: AHS 70th forum, Montreal

Corrosion Propagation Under Paint Films on Galvanized Steel: A Comparison of Phosphating and Thin Film Corrosion Pretreatment Technologies

Konrad Tarka, Annabelle Jaako, Dan Persson, Håkan Mattsson and Lars-Gunnar Johansson

Abstract The pretreatment of car bodies prior to applying the paint is crucial to promote paint adhesion and corrosion inhibition. The current phosphating pretreatments' inability to pretreat magnesium and a body with aluminum content of over 30%, as well as future environmental legislation, will force a change in the pretreatment technology. The sustainable alternative pretreatment systems do not contain heavy metals, reducing the need for post-treatment of the waste-water and sludge from the process. The change will enable more lightweight materials in car bodies resulting in CO emissions. The process baths for the environmentally friendly pretreatments do not require much energy input as they can run at ambient temperature compared to phosphating baths that are heated to 60° C. The understanding of corrosion and the role of new pretreatments increase the confidence to build safe and reliable automobiles. The environmentally friendly pretreatments in this study are commercially available zirconium oxide-based pretreatments with silane or ceramic additives. The corrosion mechanisms were studied on hot dipped galvanized steel and painted. The study was performed by exposing the samples in a cyclic corrosion test for 3 and 13 weeks. The corrosion products were analyzed using XRD, SEM/EDS, and FT-IR techniques. The corrosion products found on the samples for all pretreatments are (Zn Cl (OH) •H O) and (Zn (CO) (OH)), as well as zinc chlorides and Zn(OH) . The corrosion front of the samples pretreated with the zirconium-based pretreatment exhibited corrosion propagation in the zinc grain boundaries. The grain boundary corrosion, as opposed to the diffuse corrosion front previously reported in the literature, is believed to stem from a decoupling of the cathodic site situated in the scribe when the corrosion blister is drying up. A more local anode/cathode pair is believed to form near the corrosion front giving rise to this behavior.

K. Tarka (✉) · A. Jaako · H. Mattsson
Surface Treatment Center, Volvo Car Corporation, Gothenburg, Sweden
e-mail: Konrad.Tarka@volvocars.com

K. Tarka
Environmental Inorganic Chemistry, Chalmers University of Technology, Gothenburg, Sweden

D. Persson · L.-G. Johansson
Corrosion and Surface Technology, SwereaKIMAB AB, Kista, Sweden

I. Denbratt et al. (eds.), *Sustainable Automotive Technologies 2014*,
Lecture Notes in Mobility, DOI 10.1007/978-3-319-17999-5_15

1 Introduction

Resistance to cosmetic corrosion is imperative for minimizing costs for the customer and for the manufacturer. The dominant corrosion protective technology used by the automotive industry is based on zinc phosphating (ZnPh). The ZnPh pretreatment is deposited as zinc phosphate crystals that form a layer with a typical thickness of a few micrometers [1, 2]. Due to the use of heavy metals such as nickel and manganese in the process, as well as problems pretreating large amount of aluminum due to sludge formation, more environmentally friendly alternatives have to be considered. Recently, commercial products have emerged on the market making use of ZrO_2 conversion coatings with silane or silica additives. These coatings feature lower consumption of heavy metals, low energy consumption due to a room temperature process and the ability to pretreat large amounts of aluminum without toxic sludge buildup. The corresponding layers are typically in the 50–100 nm range [3]. In this study two commercially available thin film pretreatments and a conventional phosphating pretreatment are compared in terms of the morphology of corrosion attack. The corrosion products were characterized by FT-IR, SEM/EDS and XRD.

2 Experimental

2.1 Material, Pretreatment and Coating

Commercial grade electron discharge textured (EDT) hot dip galvanized steel with a zinc coating thickness of 7 μm was used. Sample size was 10 × 20 cm. Three different pretreatments were carried out using an industrial process during running production. Pretreatments α and β are commercially available zirconium based conversion coatings while the third pretreatment is a zinc phosphate coating (ZnPh). Subsequently, all samples were coated with a cathodic electrocoat which was applied during running production. The samples were then painted with a standard three layer automotive paint system. Before the corrosion exposures the samples were scribed, exposing the steel substrate. One sample was subjected to corrosion testing for each pretreatment (ZnPh, α and β).

2.2 Accelerated Corrosion Exposure

The samples were exposed for 3 weeks to a cyclic accelerated corrosion test developed by Volvo Car Corporation (ACTII [4]). The corrosion test operated in 168 h cycles and included cycling of temperature (25–50 °C) and humidity (70–95 % RH). During the wet periods rain was simulated three times by spraying a 0.5 wt% NaCl solution on the panels. The last “rain” in each cycle was substituted by hand spraying the samples with an aqueous solution containing 0.321 M KCl and 0.107 M NaCl.

2.3 Sample Characterization

Crystallographic characterization was performed by low angle X-ray diffraction (XRD) using a Bruker D8 instrument. The X-ray source was fixed at 2°.

Morphological and chemical characterization was performed using scanning electron microscopy (FEI Quanta 200 FEG-SEM) coupled with an energy dispersive spectrometer (Oxford X-Max). Prior to the XRD and SEM measurements the paint layers were removed using n-methylpyrrolidon (Merck, Germany) at 60–70 °C. FTIR measurements were performed using a Varian 7000 spectrometer with a UMA 500 microscope.

3 Results

3.1 Characterization with XRD

Figure 1 shows typical X-ray diffractograms measured after the corrosion test for samples subjected to the three pretreatments. Care was taken to analyze only the blistered surface. The strongest peaks are associated with metallic Zn. All three pretreatments gave strong evidence for simonkolleite ($Zn_5Cl_2(OH)_8 \cdot H_2O$). In

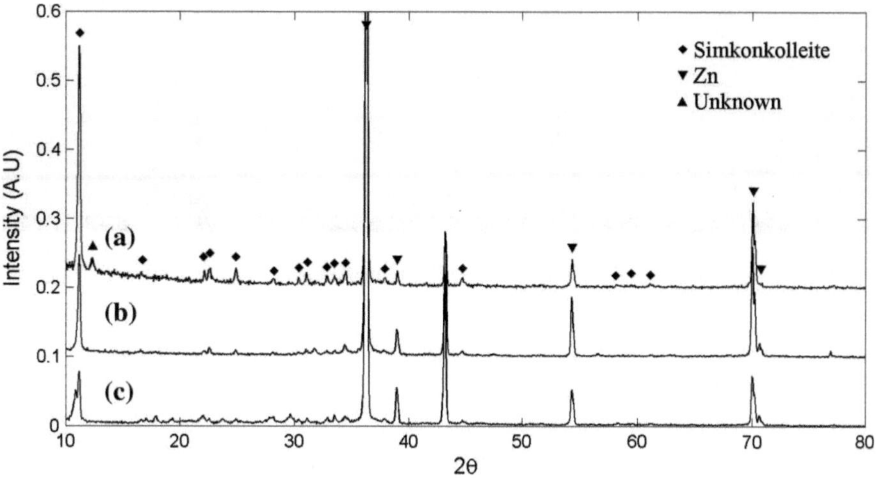

Fig. 1 X-ray diffractograms of the samples; *a*–*c* denote pretreatments α, β and ZnPh, respectively

addition, diffraction pattern c in Fig. 1 exhibited minor peaks that are attributed to the ZnPh pretreatment layer. Except for a peak at $2\theta = 12.35°$ on sample α, there were no unidentified peaks.

3.2 Characterization of the Corrosion Products with SEM/EDS

EDS data from the different regions indicated in Fig. 2 are presented in Table 1. The corresponding analysis for pretreatment β was essentially the same as for α and is not shown.

Figure 2a shows the phosphated material. The scribe is at the bottom of the micrograph and the blistered area forms a roughly semicircular area above the scribe (the hatched line indicates the extension of the blistered area). The ZnPh pretreatment layer is seen outside of the blistered area. The central parts of the blistered area were covered by the primary blister (1 and 2 in Fig. 2a) while the periphery was covered with secondary blisters (3 in Fig. 2a). In the center of the blistered area corrosion products form agglomerates consisting of <0.5 μm thick platelike crystals (not shown).

The materials subjected to the two thin film pretreatments form blistered areas exhibiting a border region with dark contrast, see Fig. 2b, c. While the morphology is similar for pretreatments α and β, the dark region is wider in the case of α. The EDT relief is preserved in the dark areas even though the zinc coating has been corroded away as indicated by the EDS analysis in Table 1. In the center of the

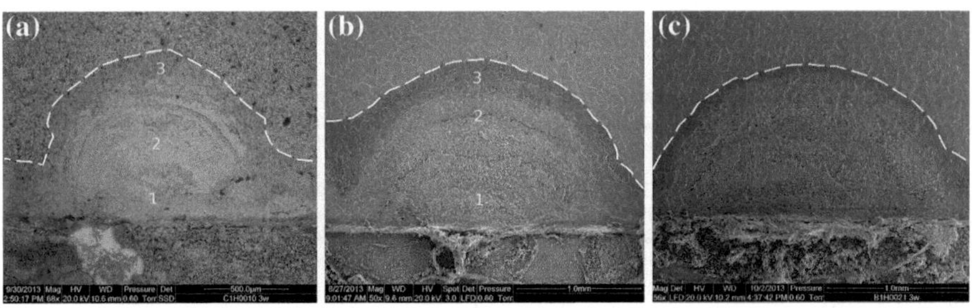

Fig. 2 Overview of blistered area; **a** shows ZnPh, **b** pretreatment α and **c** shows pretreatment β

Table 1 EDS results for the areas designated in Fig. 2

Figure 2a (ZnPh)				Figure 2b (pretreatment α)			
Area	1	2	3	Area	1	2	3
P at.%	1.3	1.1	3				
Zn/Cl	2.76	2.67	4.41	Zn/Cl	2.73	2.57	2.83
Zn/O	0.42	0.42	0.3	Zn/O	0.48	0.44	0.46

Table 2 EDS analysis of the areas indicated in Fig. 3b

Area	1	2	3	4
Zn/Cl	4.15	3.51	3.81	6.05
Zn/O	0.64	0.54	0.57	0.68

blisters formed using pretreatments α and β there are large hexagonal crystals surrounded by densely packed smaller crystals (not shown).

Figure 3 shows micrographs of the corrosion front which are typical of both thin film pretreatments α and β. The low magnification image in Fig. 3a shows the EDT textured zinc coating surface, consisting of plateaus (indicated by dotted lines) and troughs. The corrosion front is sharp at the plateaus while it is more diffuse in the troughs. Thus, in the latter regions the Zn grain boundaries are preferentially attacked. The observation of grain boundary corrosion is confirmed by the EDS chlorine map in Fig. 3c, showing penetration of chlorine into the zinc layer, along grain boundaries. The EDS analysis shows an increased Zn signal near the corrosion front compared to the center of the blister, Table 2. The Zn/Cl ratio varies from 3.51 to 6.05 (Table 2). The EDS analysis yielded similar results for pretreatment β (not shown).

The SEM image in Fig. 4 shows the periphery of the blistered area for the ZnPh pretreatment. The uncorroded zinc coating with the ZnPh coating is at the top. The bottom part shows a corroded area corresponding to the primary blister. In between there is an area with darker contrast that lacks the bright spots seen in the uncorroded area. In the intermediate region the hopeite morphology is retained. The EDS analysis of that region shows that the phosphorous concentration is similar to that in the uncorroded area (points 7–12 in Fig. 4). The Cl concentration is <0.5 at.% in the uncorroded area and increases strongly in the intermediate area and peaks in the corroded area, see Table 3.

Figure 5 shows a micrograph of the scribe for pretreatment α and a corresponding quantitative Cl map. There is a deficit of Cl in parts of the scribe, Fig. 5b.

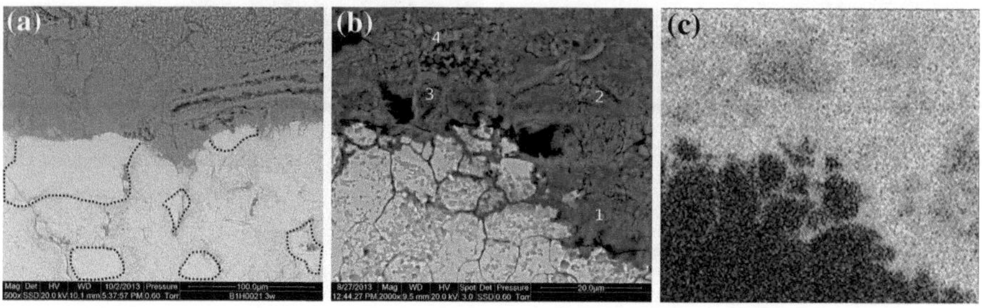

Fig. 3 Micrographs showing representative images of the corrosion front for the two thin film pretreatments α (**a**) and β (**b**). **c** Shows the Cl map corresponding to (**b**). Note that the magnification is lower in (**a**) than in (**b**)

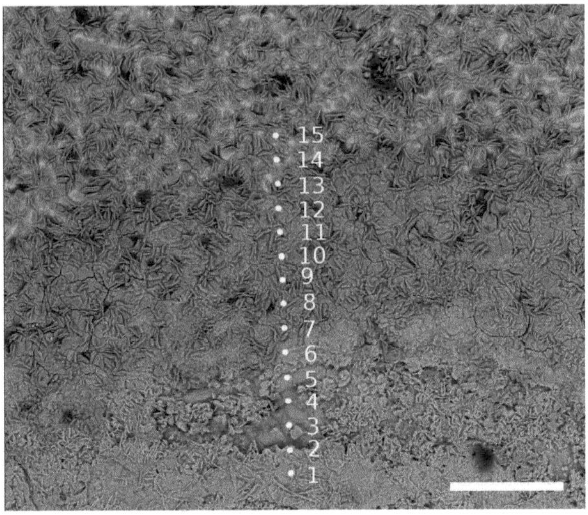

Fig. 4 Micrograph showing the corrosion front for the ZnPh pretreatment. The *numbered* points correspond to the EDS spectra obtained. The *scale bar* is 50 μm

Table 3 EDS point analysis at the positions indicated in Fig. 4

Point	1	6	12	15
P at.%	1.32	1.9	4.54	6.62
Cl at.%	6.28	5.58	4.07	0.42
Zn/Cl	3.66	3.75	4.4	41.12
Zn/O	0.62	0.49	0.42	0.33

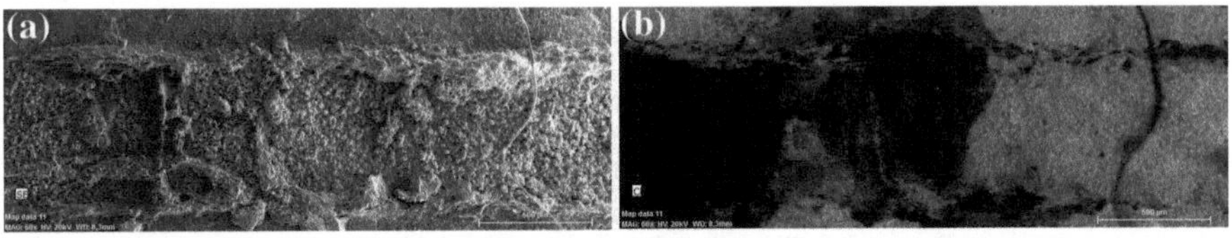

Fig. 5 Micrograph of the scribe on the ZnPh sample (**a**) and Cl mapping of the same area (**b**)

3.3 Characterization with FT-IR

Figure 6A shows FT-IR spectra from sample α (the results for sample β were similar and are not shown) while Fig. 6B shows the corresponding analysis of the ZnPh sample. The spectra obtained behind the corrosion front for samples α and β, (Fig. 6Ab) and ZnPh (Fig. 6Bc) are interpreted as simonkolleite based on the bands at 1040, 911, and 748 cm^{-1}. The thin film pretreated samples show a signal that

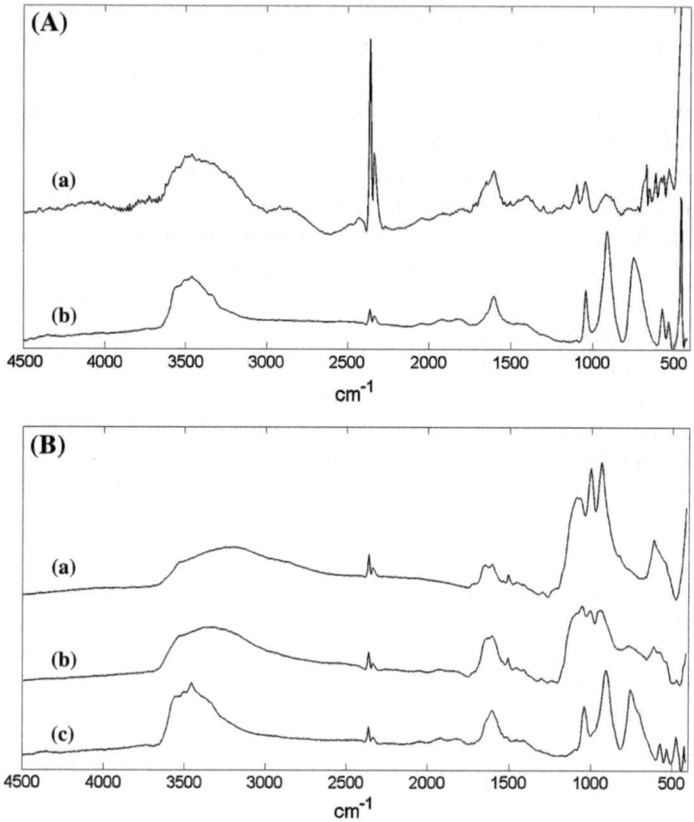

Fig. 6 FT-IR spectra from sample α (**A**) and ZnPh (**B**). In diagram **A** the *a* and *b* spectra are at and behind the corrosion front, respectively. In diagram **B** spectrum *a* is ahead of, *b* is at and *c* is behind the corrosion front

suggests the presence of $Zn(OH)_2$ at the corrosion front due to the bands at 1096 and 1046 cm^{-1}, Fig. 6Aa [5]. The spectra obtained from the phosphated sample, Fig. 6B, shows a gradual change from simonkolleite to hopeite [6] as we move across the corrosion front into the uncorroded area, which is consistent with the SEM images in Fig. 2a.

4 Discussion

In accordance with previous studies [2, 7], simonkolleite is the main crystalline corrosion product for all pretreatments. The morphology of simonkolleite on the thin film pretreated samples is different from that on the phosphated material. It is suggested that the formation of larger crystals on the former may be due to a lower degree of supersaturation of the electrolyte. The Zn/Cl ratios measured in the blistered areas are higher than the corresponding ratio for simonkolleite which is 2.5. This implies that the volume analyzed by SEM/EDX also includes other substances, e.g. zinc hydroxide or metallic Zn. Hydrozincite ($Zn_5(CO_3)_2(OH)_6$) which has been reported to form under similar conditions, is likely to be found near the scribe as this is the most probable point of entry for CO_2. This explains the lack of a carbonate band in the FT-IR spectra which were obtained near the corrosion front.

Based on the SEM/EDX (Figs. 2 and 3) and FT-IR (Fig. 6) analysis, the corrosion blisters formed on the thin film pretreated materials are interpreted in terms of spatially separated anodic and cathodic areas. Thus, the Cl deficient area in the scribe is suggested to be the main cathodic area (Fig. 5), at least during the wet part of the cycle. Similarly, the chloride enrichment at the corrosion front indicates that the area in question is anodically active, see Fig. 3c.

On the thin film pretreated material the grain boundaries in the zinc coating corrode preferentially, producing an archipelago-like structure. This is visible primarily in the interplateau areas, presumably because of the smaller zinc grain size in those regions, being due to recrystallization during the texturing process [9]. The detection of $Zn(OH)_2$ at the corrosion front by FT-IR indicates the presence of a cathodic area close to the corrosion front. It is suggested that a secondary cathode forms during the low humidity portion of the exposure cycle as the corrosion current decouples from the main cathode in the scribe. This is in accordance with the observation of corrosion of the exposed steel in the scribe [7]. The corrosion front on the ZnPh samples is covered with phosphate crystals and is hence more difficult to analyze compared to the thin film treated materials. The presence of the phosphate layer indicates that pH is in the range 4–11 [8]. The results imply that the overall mechanism for paint delamination is anodic undermining [7] which seems to be the case for all pretreatments.

In conclusion, the corrosion mechanism for all investigated pretreatments appears to be similar. However, the differences in blister morphology and corrosion product morphology imply that the environment in the corrosion blisters is not the same for the thin film pretreatments as for the phosphated material.

References

1. Narayanan TSNS (2005) Rev Adv Mater Sci 9:130–177
2. Sabata A (1989) PhD Thesis T-3777, Colorado School of Mines Golden, Colorado
3. Adhikari S, Unocic KA, Zhai Y, Frankel GS, Zimmerman J, Fristad W (2011) Electrochim Acta 56(4):1912–1924
4. Volvo Car Corporation Corporate Standard VCS 1027,1449, issue 1
5. Secco EA (1967) Can J Chem 45:585–588
6. Stoch A, Paluszkiewicz C, Długoń E (1999) J Mol Struct 511–512:295–299
7. Amirudin A, Thierry D (1996) Prog Org Coatings 46(1):162–175
8. van Ooij WJ, Sabata A (1989) Surf Coatings Technol 39(40):667–674
9. Mataigne J-M, Vaché V, Repoux M (2009) Rev Métallurgie 106(1):41–47

Lightweight Design of Composite Sandwich Structures

H. Bansemir

Abstract Since a long time lightweight sandwich structures were used in the design of aircraft components. But the more recent use of composites allows even lighter components with outstanding special features. Beginning in the early 60 s sandwich rotor blades (e.g. for the helicopter BO 105) were designed with composite face skins and mainly foam for the support of the skins in order to prevent the skins from local buckling failure. The design of the BO 105 cross section, the blade cross section of the EMMEN wind tunnel, the DNW wind tunnel blade are shown in the presentation of this paper. The foam support allows an improved local stability of the skins. The stability strength is dependent on the longitudinal stiffness of the face skins and the transverse and shear stiffness of the supporting foam. Carbon- or Glass-fibers are used for the blade skins. Extreme light and stiff sandwich carbon composite designs are used since the early 70 s for space satellite solar arrays. The design consists of carbon face skins which are filament winded open nets. The solar cells are isolated to the carbon filaments by thin Capton foils. For the needed high bending stiffness, extremely high modulus carbon fibers are used. Again wrinkling strength is the critical failure mode, due to high dynamic acoustic noise produced by the space-craft booster. The noise level and thus the load can be reduced with the help of "Helmholtz-Resonators" integrated in the multifunctional design. This kind of sandwich design can also be used for light weight fuselage structures. Multifunctional features, such as damping elements can be included in the design.

1 Introduction

The outstanding behaviour of fibre composites allows new sandwich designs with low weight and other special properties. The sandwich structures of wind energy rotor blades use often low cost foams. In order to reduce weight, low density

H. Bansemir (✉)
Consulting for Composites, Light Weight Structures and Knowledge Management, Geitnerweg 4, 81825 Munich, Germany
e-mail: horst@bansemir.eu

I. Denbratt et al. (eds.), *Sustainable Automotive Technologies 2014*,
Lecture Notes in Mobility, DOI 10.1007/978-3-319-17999-5_16

Aluminium-, Kevlar- or paper-honeycomb cores are used. The combination of Aluminium honeycombs with carbon fibres can produce failures because of their different special electrical characteristics.

The face sheet material can be of very different art. As an example, the light weight wings of the communication satellites have panels with an open net of high modulus carbon fibres. This arrangement allows a very low and stiff design.

The first helicopter composite sandwich blades in serial production were used for the BO 105 in the year 1967. Glass fibre fabric and unidirectional material is used to ensure a good dynamic behaviour and an optimal fatigue resistance.

Newer designs of helicopter blades use carbon fiber fabrics for an improvement of the torsional stiffness.

In the early 80 s MBB (now Airbus Helicopters) studied and designed blades for wind energy converters as seen in Fig. 1. In the Fig. 3 wind energy converters with one and two blades are shown. Carbon fiber material was used for a high bending stiffness of the blades. As the price of carbon is much higher than for glass material the present design uses mainly glass fibers. The rotor of AEOLUS II (Fig. 3) has a maximum power output regarding two blades with a length of 40 m of 3 MW. The power output can be increased to 5 MW and possibly to 7 MW, regarding 3 blades with a length of about 50 m. The helicopter industry is improving the fuselage structures introducing additional features such as the ability to survive high crash loads.

Fig. 1 The wind-energy converter-park with Aeolus II and MON 50 (Bansemir and Pfeifer 1980) [2]

2 Blades for Composite Rotor Systems and Wings

The composite wings of the early German gliders were introduced with the FS 24 Phoenix of the Stuttgart Akaflieg student group in 1957. In the early 60 s composite rotor blades were designed for rotor systems of helicopters BO 103 and BO 105. The BO 105 flew in 1967 for the first time. With the design, elastic hinges could be incorporated and thus simplified the rotor system. The design life could be extended because of the improved fatigue behavior. In the late 70 s the metallic blades of the wind channel EMMEN failed. Composite rotor blades were designed with carbon composites, the cross section of the blade is shown in Fig. 2. At the same time, the wind channel DNW was developed with sandwich blades using carbon fibers and epoxy resin. The cross section is shown in Fig. 2. In Fig. 3 the development of blades for wind energy converters are shown. The WEA and the Monopteros-30 and 50 had only one blade, constructed with carbon composites. The WEC and AEOLUS II used two blades, designed also with carbon and glass composites. Recently big energy converters (up to 5 MW) are used for huge established wind-parks. Glass fibers are used, because of the lower material price. The bended MON 50 rotor blade is shown in the Fig. 1. The critical failure mode is local stability (wrinkling). In Fig. 4 the failure of a sandwich specimen in a four point bending test rig is shown. With the given formula, the wrinkling stress can be calculated, taking into account the shear modulus G of the foam, the longitudinal face sheet Young's modulus E and E of the foam.

The factor of 0.825 is the optimum theoretical value. As a practical value, 0.5 is usually taken, regarding manufacturing defaults and waviness of the composite layers.

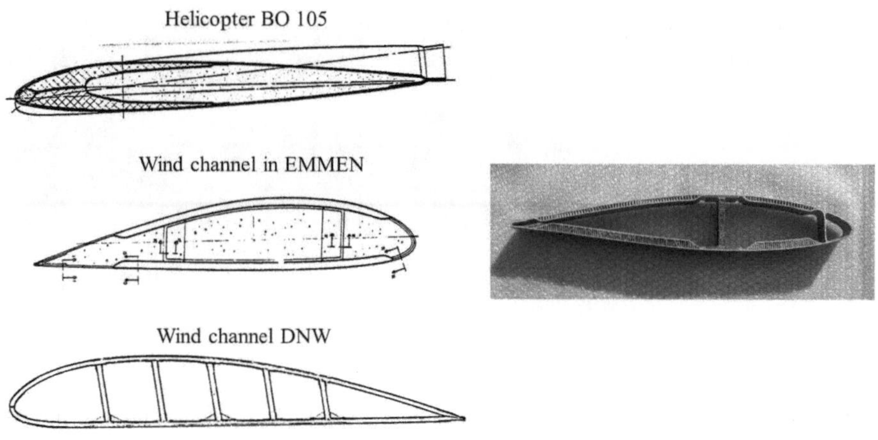

Fig. 2 Cross-sections of the helicopter blades BO 105, wind-channels EMMEN and DNW and horizontal stabilizer BO 108 (Bansemir and Pfeifer1980) [2]

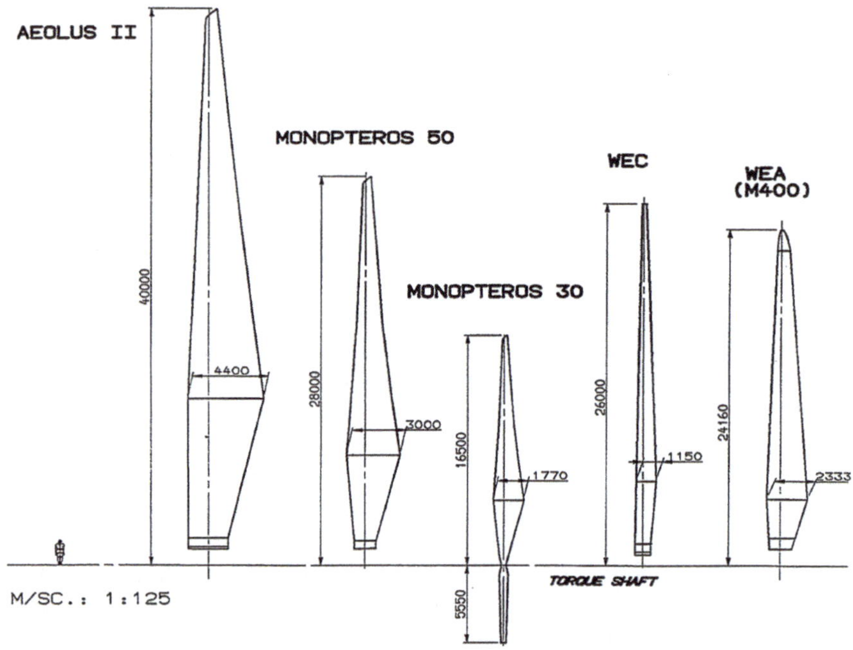

POWER DIAMETER	3000 kW 80 m	650 kW 56 m	200 kW 33 m	285 kW 52 m	450 kW 48 m

Fig. 3 The development of the different rotor blades at MBB in the Years 1980–992

$$\sigma_{kn} = 0.825 \sqrt[3]{E_H E_K G_K}$$

Fig. 4 The four point bending test of a sandwich beam and the formula for the wrinkling strength [2]

In Fig. 2 on the right hand side a cross section of a horizontal stabilizer of the BO 108 is shown. The extreme light design includes honeycomb core, in order to reinforce the outer skin and prevent cross section deformations.

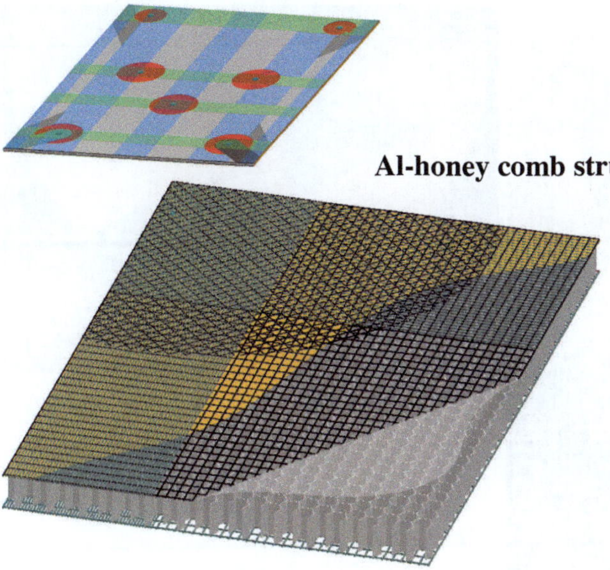

Fig. 5 Al-honey comb sandwich structure with open-net carbon fibre composite face sheets for satellite solar array (Bansemir ICSAT 2011) [1]

3 Carbon Composite Sandwich Structures for Space Satellites

Special features such as thermo-elastic stability and high stiffness are needed for the construction of satellite structures. The composite sandwich design shown in Fig. 5 has open net face sheets and light weight Al honey combs as core material. The basic high modulus carbon fibers are laid up in a 0/90 degree open net. Additional layers are used to reinforce the common face sheets. As seen in Fig. 5, load introduction and high loaded areas have additional carbon reinforcements. The shown face sheets are filament winded with epoxy resin and cured at high temperature. The filament winding process allows a very light face sheet construction. The open net design allows the evacuation of air when entering the high altitude space area. The Al honey combs are perforated for a better air fluctuation. The air fluctuation is an important feature for space satellites. Special design elements are used for the load introduction hold down points.

4 Al-Honey Comb Structure with CFC Face Sheets

The critical load situation is acoustic noise, produced by the space rocket. From these loads bending moments result and the failure mode is wrinkling strength (Fig. 6).

Fig. 6 The scientific satellites Venus (*left*)- and, Mars (*right*)-express (Haider and Bansemir)

The formula for the wrinkling strength takes into account the anisotropic core material. The local face sheet bending stiffness B should be used. The wrinkling strength formula contains the

$$B = \frac{E_f t_f^3}{\left(1 - \nu_{xy}\nu_{yx}\right)12}$$

$$\sigma_{WR} = k\frac{\sqrt{12}}{t_f}\sqrt{\frac{E_c}{h_c}B}$$

height h of the core, the thickness of the face sheet t, the Young's modulus E and the factor k. Usually a factor of 0.33, taking manufacturing defects into account, should be used.

In Fig. 7 a recently developed solar array for a communication satellite Eurostar 3000 is shown.

5 Multifunctional Sandwich Fuselage Helicopter Structures

In Fig. 8 the military transport helicopter NH 90 is shown. The centre section of the fuselage is specially designed for bearing high crash loads.

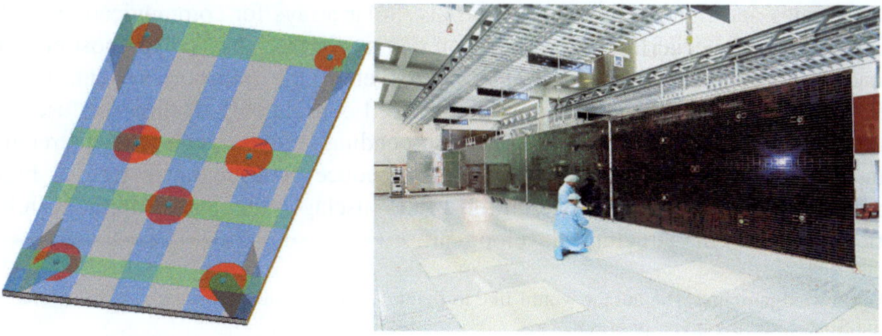

- **Modular design: 3-5 panels, 2 sizes, Si / GaAs**
- **Intelsat-X measured data:**
- **11.7 kW EOL (End of Life)**
- **(5 large panels, 4 Si + 1 GaAs)**

  length 22 m (1 wing)

  mass 280 kg (2 wings)

Fig. 7 The recent communication satellite Eurostar 3000 (Haider 2013)

Fig. 8 The centre section of the fuselage of the Helicopter NH 90 (Frommlet 1997)

For future developments, multiple functions can be installed in sandwich structures, such as crash survivability, acoustic improved behaviour, structural damping and improved possibility of damage tolerance and repair. The all composite fuselage of the military NH 90 is an important example for future light weight transport vehicles.

6 Conclusion/Summary

Sandwich structures are widely used since a long time. New design possibilities and multifunctional structures can be introduced by using composite materials. Blade structures for helicopters and wind energy converters are developed with the help of composites and light weight core materials. Solar arrays for communication satellites are using special light weight carbon composite face sheets. The most critical load is acoustic noise produced by the rocket of the transport system. Using Helmholtz Resonators the loads can be reduced considerably. Helicopter fuselage structures should have a high torsional and bending stiffness, high fatigue strength and sustain high crash loads. Measures to reduce acoustic noise have a high priority in order to improve the comfort of the fuselage. Structural damping can be incorporated in the structural development.

Acknowledgements The author would like to thank "Airbus Helicopters" and "Airbus Defence and Space" for the given photos.

References

1. Bansemir H (2011) Design of basic structural composite elements. In: Proceedings of ICSAT2011, Greenville, USA
2. Bansemir H, Pfeifer K(MBB) (1980) Stress analysis and test philosophy for wind energy converter blades. In: Fourth meeting of rotor-experts, with respect to fatigue design problems, Stockholm, 21–22 April, 1980
3. Schürmann H (2004) Konstruieren mit Faser-Kunststoff-Verbunden, Springer-Verlag Berlin, Heidelberg, New York, ISBN 3-540-40283-7
4. Bansemir H, Haider O (1998) Fiber composite structures for space applications—recent and future developments, Cryogenics, 38(1):51-59
5. Haider O, Bansemir H (1996) Solargeneratorstrukturen in Faserverbundbauweise für Nachrichtensatelliten, Deutsche Gesellschaft für Luft- und Raumfahrt—Lilienthal-Oberth e. V., Tagung des DGLR-Fachbereichs T1 Strukturen, München, 27./28. November 1996
6. Bansemir H, Bauer M, Haider O (2001) Eurocopter Deutschland GmbH, 81663 München: Großflächige Solargeneratoren in Faserverbundbauweise für Nachrichtensatelliten, Deutscher Luft- und Raumfahrtkongress 2001, 17. bis 20. September 2001 in Hamburg „Satellitentechnologie"
7. Bansemir H, Haider O (1991) strength and deformation behavior of kevlar and carbon composite sandwich structures for space applications, pp 24–26
8. Frommlet H (1997) NH90, „Ein Europäisches Hubschrauber-Programm", Vortrag anlässlich der DGLR Jahrestagung, 14.-17. Oktober 1997 in München

Part VI
Alternative and Renewable Fuels

Effects of Natural Gas Percentage on Performance and Emissions of a Natural Gas/Diesel Dual-Fuel Engine

Zhiqin Jia and Ingemar Denbratt

Abstract Due to rising costs of conventional fossil fuels, and increasingly stringent limits on emissions (especially "greenhouse gases"), use of cleaner, cheaper gaseous fuels in internal combustion engines is expected to increase in the future. A popular application is the operation of heavy-duty diesel engines using combinations of compressed natural gas (CNG), supplied with the intake air and diesel injected to initiate combustion. Extensive efforts have already been made in both industry and academia to minimize pollutant emissions and maintain diesel-equivalent performance in this dual-fuel mode, but further knowledge of effects of fundamental parameters is required to optimize the combustion. Thus, this paper presents an experimental investigation of the influence of the CNG to diesel fuel ratio on the performance (effective expansion ratio, pressure and heat release rates) and emissions (HC, CO, NO_x and CO_2) from a CNG/diesel dual-fuel engine operating under varying load conditions but constant engine speed.

1 Introduction

The automotive industry is still highly dependent on conventional fossil fuels globally. However, due to increases in their prices and concern about environmental pollution there is growing interest in the use of alternative fuels. Among these alternative fuels, compressed natural gas (CNG) is a popular choice because of its relatively clean combustion and global availability at attractive prices. Furthermore, with minor modifications heavy-duty diesel engines can operate using mixtures of CNG and diesel. Considerable efforts have been made to assess and improve the performance and emissions of engines operating in such dual-fuel combustion mode recently [1–5], but further optimization is required.

Z. Jia (✉) · I. Denbratt
Chalmers University of Technology, Gothenburg, Sweden
e-mail: zhiqin@chalmers.se

I. Denbratt et al. (eds.), *Sustainable Automotive Technologies 2014*,
Lecture Notes in Mobility, DOI 10.1007/978-3-319-17999-5_17

Table 1 Basic engine specification

Engine type	Volvo D12C
Bore	131 mm
Stroke	150 mm
Connecting rod	260 mm
Compression ratio	17:1
Engine speed	1500 rpm
No. of nozzle orifices	7
Injector nozzle diameter	0.225 mm
SOI electric signal	6 BTDC

In heavy duty dual-fuel operation the CNG is injected into the intake manifold and mixed with air before being inducted into the cylinder. The mixture enters the cylinder during the intake stroke and is compressed during the compression stroke, as in a conventional diesel engine. A certain amount of diesel oil is injected in the cylinder at the end of compression stroke to initiate the gas-air mixture combustion. Compared to CNG, diesel oil has relative high cetane number which can auto-ignite under compression. While CNG has relative high octane number which prevent CNG from auto-igniting during the compression stroke.

In the presented work we conducted tests with a single-cylinder version of the Volvo D12C diesel engine, with a displaced volume of 2.02 L. Mixtures of CNG and diesel oil with various mass ratios were used to examine their effects on engine performance and emissions.

2 Experimental Setup and Methodology

2.1 The Single-Cylinder Engine

The single-cylinder variant of the Volvo heavy duty D12C engine used in the experimental studies was equipped with a four-valve cylinder head, a centrally positioned diesel injector and a quiescent combustion system. The major engine specifications are listed in Table 1. In addition the engine was equipped with oil, water and air conditioning systems, and a short route EGR system with cooling from the outlet of the engine. The inlet air was supercharged by a screw-compressor, dehydrated in a dryer and temperature-regulated.

A schematic layout of the experimental setup is shown in Fig. 1. The engine was modified to enable dual-fuel combustion by adding a gas injection system to the intake manifold. CNG was obtained from a local distribution network. Before entering the engine cylinder, the CNG was passed through a pressure regulator to reduce its pressure from 200 to 14 bar and then through a gas flow meter. Following the injection signal the gaseous fuel was injected into the engine's intake and mixed with the intake air. The gas and diesel fuel control systems were completely

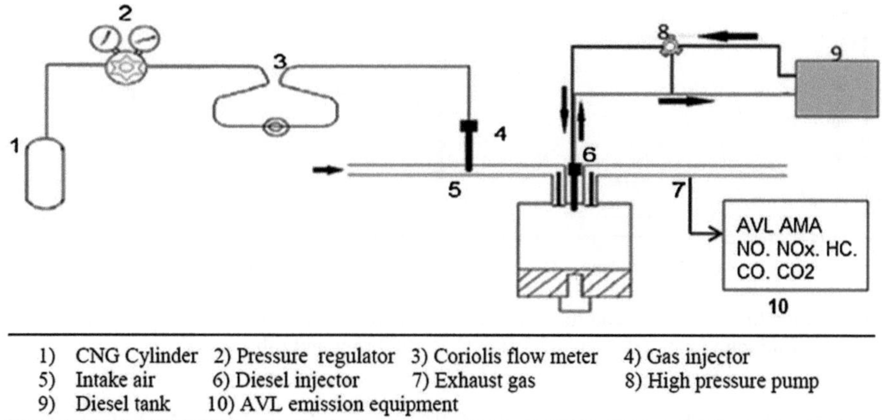

Fig. 1 Schematic layout of the experimental setup

independent from each other. A typical rail-type fuel injection system was used to inject the diesel.

2.2 Methodology

The experiments were performed with an engine speed of 1500 rpm, an SOI of 6 CAD BTDC (electrically signaled), an injection pressure of 2000 bar for the diesel oil and at load points of both 25 % (6 bar BMEP) and 50 % (12 bar BMEP). The mass ratio of CNG was varied from 0 to ~90 % under each of the load conditions. The engine's performance was characterized by measuring the in-cylinder pressure, heat release rate, fuel consumption and gas temperature. Engine emissions (NO_x, HC, CO and CO_2) were also recorded. The NG mass ratio was calculated using the following equation:

$$x = \frac{\dot{m}_{NG}}{\dot{m}_{NG} + \dot{m}_{Diesel}} * 100\%$$

Here, $\dot{m}_{NG}$ and $\dot{m}_{Diesel}$ are the mass flow rate of the NG and diesel, respectively.

3 Results and Discussion

As already mentioned, the objective of the study was to investigate effects of the NG to diesel ratio on the performance and emissions of a DI diesel engine operating in dual-fuel mode, at constant engine speed (1500 rpm) and two loads (6 and 12 bar

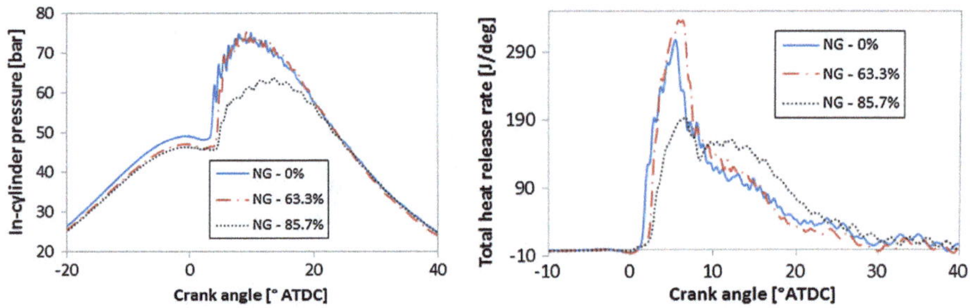

Fig. 2 In-cylinder pressure and total heat release rate traces obtained from tests with pure diesel and diesel mixtures with 63.3 and 87.5 % NG at 1500 rpm and 6 bar BMEP

BMEP). The results, presented in the following sections, do not include measurements of soot emissions, because they were very low (non-detectable in some cases). This is consistent with expectations because soot emissions are not generally problematic in CNG/diesel combustion.

3.1 Cylinder Pressure and Total Heat Release Rate Traces

Figure 2 shows in-cylinder pressure and total heat release rate traces obtained from tests at the light load cases (6 bar BMEP) at 1500 rpm. The traces in the left panel show that the in-cylinder pressure increased much more rapidly after ignition in tests with both pure diesel and 63.3 % NG than with 83.7 % NG (which also resulted in a lower peak pressure). The heat release rate curves in the panel indicate that the combustion with pure diesel is predominantly premixed, with a minor contribution from diffusion combustion. A possible explanation for this is that less fuel is needed at light load, which can be premixed well during the ignition delay time. Therefore, use of 63.3 % NG provides little advantage under these conditions. Furthermore, increasing the NG percentage to 85.7 % results in a slower combustion rate and thus a lower and later pressure peak than pure diesel. The lower burning rate may be due to deterioration of the combustion since less diesel fuel is available to ignite the mixture. The heat release rate trace obtained with 85.7 % NG shows that the combustion duration was longer than with pure diesel, indicating that combustion was less efficient. The traces also show that the ignition delay was longer with both dual-fuel mixtures than with pure diesel.

As shown in Fig. 3, at the high load (12 bar BMEP), the in-cylinder pressure after ignition increased more rapidly and there was a higher peak pressure with 69 % NG than with pure diesel, possibly due to greater evaporation and mixing of the pilot diesel spray during the pre-ignition period and higher gas concentrations surrounding the spray [5]. The heat release rate traces in the figure indicate that with both pure diesel and 69 % NG the combustion processes included two-phase combustion (pre-mixed combustion and diffusion-controlled/pre-mixed NG/air

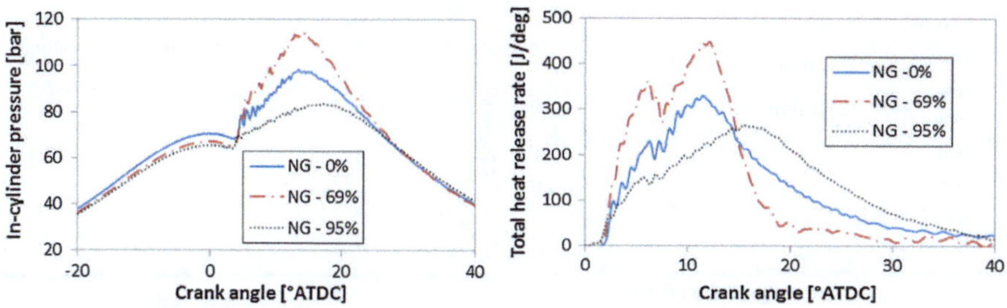

Fig. 3 In-cylinder pressure and total heat release rate traces obtained from tests with pure diesel and diesel mixture with 69 and 95 % NG at 1500 rpm and 12 bar BMEP

mixture combustion). However, the 69 % NG mixture burned more rapidly and resulted in an overall higher heat release due to more fast burning of pre-mixed NG/ air during the second phase combustion than diffusion-controlled combustion of pure diesel case. Increasing the load from 6 to 12 bar BMEP raises the temperature in the cylinder, and thus the flame propagation speed. Therefore, dual-fuel combustion with 69 % NG at the higher load was much more efficient than with 63.3 % NG at the lower load. However, with 95 % NG the combustion was less efficient, manifested in a slower heat release rate (Fig. 3), presumably due to the small quantity of pilot diesel. Thus, the pressure increased more slowly and the peak pressure was lower than in tests with both pure diesel and 69 % NG.

As shown in Fig. 4, at the high load the effective expansion ratio (EER) [6] increased with increases in the NG ratio to ca. 83 %, but declined with further increases to a lower level than in tests with pure diesel.

$$EER = \frac{\int_{SOC}^{EOC} HHR(\theta) \cdot ER(\theta) \cdot d\theta}{\int_{SOC}^{EOC} HHR(\theta) \cdot d\theta}$$

where θ, HHR(θ) and ER(θ) are the crank angle, the instantaneous heat release rate and the instantaneous expansion ratio.

In all the cases discussed above, under both low and high load conditions, the in-cylinder pressure decreased with increases in the NG percentage during the

Fig. 4 EER as function of natural gas mass ratio at 1500 rpm and 12 bar BMEP

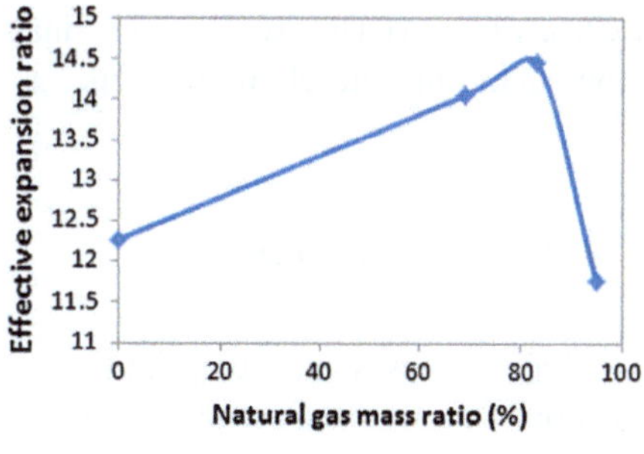

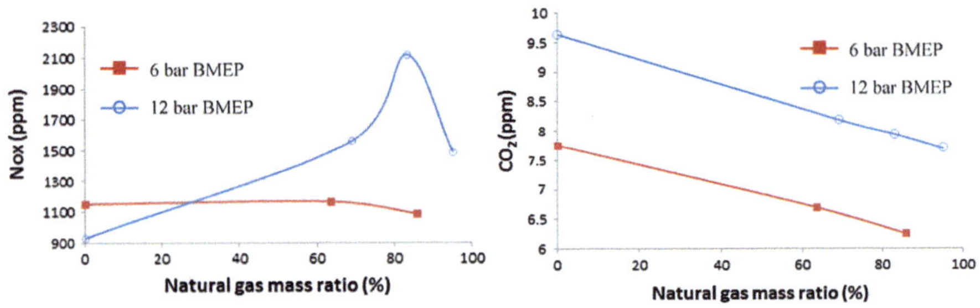

Fig. 5 Nitric oxide and carbon dioxide as function of natural gas mass ratio at 1500 rpm for 6 and 12 bar BMEP

compression stroke, presumably due to reductions of the mixture heat capacity ratio γ with increases in the NG mass ratio.

3.2 Nitrogen Oxide Emissions

Nitrogen oxide (NO_x) emissions are influenced by numerous variables, including the combustion temperature, oxygen concentration and combustion duration [7–9]. In our tests NO_x emissions were higher at the low load than at the high load when using pure diesel (Fig. 5, left panel). Furthermore, they slightly decreased with increases in the NG percentage at the low load, possibly due to deterioration of the combustion process and reduction of the charge temperature associated with reduction of the pilot diesel quantity. At high load NOx emissions were higher than at low load with all the tested dual fuel ratios, and increased with increases in the NG percentage. This is presumably because NG-air mixtures are richer at high load, thus flame propagation is faster and the charge temperature higher.

3.3 CO_2 Emissions

CO_2 emissions declined with increases in the NG percentage at both loads, presumably because NG has a lower C/H ratio than diesel. Even including unburned hydrocarbons (UHC, see below), dual-fuel combustion yielded lower CO_2 emissions than pure diesel combustion.

3.4 UHC Emissions

Several processes may contribute to UHC emissions, including inefficient combustion, wall wetting, flame quenching near cold walls and fuel passing through the

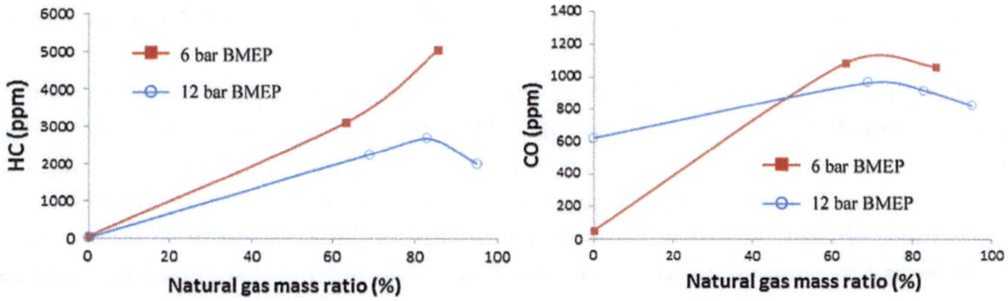

Fig. 6 Unburned hydrocarbon and carbon monoxide as function of the natural gas mass ratio at 1500 rpm for 6 and 12 bar BMEP

engine without burning during valve timing overlaps [7, 8, 10]. In our tests, use of the dual-fuel consistently resulted in higher UHC emission than use of pure diesel, and higher UHC emissions at light load than at high load (Fig. 6, left panel). This can be explained by the leaner mixtures, less efficient and slower combustion, and lower charge temperatures (as detailed above), all of which can contribute to flame extinction (bulk quenching), resulting in some of the fuel mixture reaching the exhaust without burning. At high load UHC emissions increased with increases in the NG percentage up to ca. 83 % NG then declined. However, they were consistently lower at high load than at low load, presumably because of the more efficient combustion, higher charge temperature, faster burning and (hence) more efficient HC oxidization.

3.5 CO Emissions

As shown in Fig. 6 (right panel), CO emissions were lower at light load than at high load when using pure diesel, but higher at low load in the tested dual-fuel cases, presumably due to the leaner mixtures, less efficient combustion, lower charge temperatures and (hence) impairment of CO oxidation. In general, use of dual-fuel generally led to higher CO emissions than use of pure diesel.

4 Conclusions

Our results indicate that when using dual (NG and diesel) fuel in a heavy-duty diesel engine at high load increasing the NG/diesel mass ratio up to a threshold level will increase the peak pressure and reduce the combustion duration. However, further increases will reduce the peak pressure and prolong combustion. We also found that NOx, UHC and CO emissions were higher in dual-fuel combustion mode, although CO_2 emissions decreased with increases in the NG/diesel mass ratio.

At light load use of the dual fuel resulted in a similar peak pressure to pure diesel up to a threshold NG/diesel mass ratio. However, further increases resulted in a reduction in peak pressure and prolonged combustion. NO_x emissions slightly decreased, while UHC and CO emissions increased with increases in the NG/diesel mass ratio. Overall, the results show that careful modulation of the NG/diesel ratio in response to changes in load (and other key parameters) is required to optimize dual-fuel combustion in heavy-duty diesel engines.

Acknowledgments We thank Dr. Savo Gjirja, Eugenio De Benito and Anders Mattsson for help with the experiments and the Swedish Energy Agency for funding.

References

1. Mustafi NN, Raine RR, Verhelst S (2013) Combustion and emissions characteristics of a dual-fuel engine operated on alternative gaseous fuels. Fuel 109:669–678
2. Mustafi NN, Raine RR (2008) A study of the emissions of a dual fuel engine operating with alternative gaseous fuels. SAE paper no 2008-01-1394
3. Papagiannakis RG, Hountalas DT (2003) Experimental investigation concerning the effect of natural gas percentage on performance and emissions of a DI dual fuel diesel engine. Appl Therm Eng 23:353–365
4. Stelmasiak Z (2002) The Impact of gas-air composition on combustion parameters of dual fuel engines fed CNG. SAE paper no 2002-01-2235
5. Papagiannakis RG, Hountalas DT, Rakopoulos CD, Rakopoulos DC (2008) Combustion and performance characteristics of a DI diesel engine operating from low to high natural gas supplement ratios at various operating conditions. SAE paper no 2008-01-1392
6. Stanton DW (2013) Systematic development of highly efficient and clean engines to meet future commercial vehicle greenhouse gas regulations. SAE paper no 2013-01-2421
7. Heywood JB (1988) Internal combustion engine fundamentals. McGraw-Hill, New York
8. Karim GA (1980) A review of combustion processes in the dual fuel engine—the dual fuel engine—the gas diesel engine. Prog Energy Combust Sci 6:277–285
9. Karim GA (2003) Combustion in gas fueled compression: ignition engines of the dual fuel type. ASME J Eng Gas Turbines Power 124:827–836
10. Krishnan SR, Biruduganti M, Mo Y, Bell SR, Midkiff KC (2002) Performance and heat release analysis of a pilot-ignited natural gas engine. Int J Eng Res 3(3):171–184

Investigation of Cold Start Operability, Long Term Impact, Emissions and Fuel Consumption of HVO Fuel in Volvo Car Engines and Vehicles

M. Johansson, L. Jacobsson and H. Simonsen

Abstract This study includes both engine and vehicle tests where HVO (hydrogenated vegetable oil) fuel was compared with diesel fuels. Cold start operability, long term impact on materials, emissions and fuel consumption were investigated. For the cold start tests the vehicle was cooled down for 10 h to selected temperatures and the time to start the engine was measured down to −27 °C. The tests showed that starting time was below 2 s down to −20 °C and there was no difference between the fuels. One car was operating 100,000 km on HVO fuel and the long term exposure showed neither negative impact on the materials nor high fuel dilution in the engine oil. Emission measurements were conducted in both test bed engine and vehicle. The test bed measurement showed that engine-out PM reduction is ~60 %, the NO_x emissions are similar and the specific fuel consumption is 4 % lower for the HVO fuel in comparison with European diesel fuel. The tailpipe CO_2, CO and HC emissions were reduced in the vehicle test by 6, 80 and 65 % respectively for the HVO fuel as a consequence of the higher hydrogen to carbon ratio.

1 Introduction

Drivers for research in alternative fuels are for example energy security and reduction of harmful emissions and greenhouse gases. HVO (hydrogenated vegetable oil) is a renewable fuel to be used in compression ignition engines. It is a paraffinic fuel produced from vegetable oil or animal fats. Moreover it has high hydrogen to carbon ratio and is therefore beneficial for reduction of CO_2 emissions.

HVO is a paraffinic fuel without aromatics and consequently features a high cetane number. In contrast to FAME (Fatty Acid Methyl Ester) it neither contains saturated monoglucerides nor sterol glucosides, which could both cause clogged fuel filters. Another aspect with FAME fuels is the poor cold flow properties which are improved by isomerization and/or cracking optimization for HVO fuel [1].

M. Johansson (✉) · L. Jacobsson · H. Simonsen
Volvo Car Corporation, Gothenburg, Sweden
e-mail: monica.fuel.johansson@volvocars.com

I. Denbratt et al. (eds.), *Sustainable Automotive Technologies 2014*,
Lecture Notes in Mobility, DOI 10.1007/978-3-319-17999-5_18

The risk of fuel dilution in engine oil is lower for fuels with lower distillation intervals. HVO fuel has a final boiling point of 300 °C and fuels containing FAME above 350 °C. Then the expected fuel dilution for HVO fuel would be lower than FAME blends.

This investigation includes both test bed engine and vehicle test where HVO fuel was compared with different diesel fuels. In vehicles cold start operability and long term impact on materials were studied. Emissions were measured in both engine and vehicle. All tests were conducted on Volvo 5-cylinder diesel engine.

2 Fuels and Methodology

The experiments were divided into cold start tests, durability, emission and fuel consumption tests.

2.1 Fuels

The fuels used in the study were HVO (Hydrogenated vegetable oil) and different diesel fuels. The fuels are specified in Table 1.

2.2 Methodology

HVO fuel was compared to different diesel fuels. The tests included in the study were:

- Vehicle cold start
- Vehicle durability
- Test bed emissions
- Vehicle emissions

In the cold start test the vehicle was cooled down for 10 h to selected temperatures and the time to start the engine was measured down to −25 °C. HVO was compared to EN590 B10.

The durability test featured a vehicle operating as a driving school car (mostly city driving) running on HVO fuel for 100,000 km. After the test, the engine, fuel tank and aftertreatment system was disassembled in order to detect long term effects.

The engine tests were performed at 1500, 2000 and 2500 rpm speed at different loads. Emissions of soot, NO_x and fuel consumption were measured. HVO was compared to EN590 B10.

In the vehicle emission tests the vehicle was operated in the NEDC cycle and emissions of CO, HC, NO_x, PM, CO_2 and fuel consumption were measured. HVO was compared to Euro 5 diesel.

Table 1 Fuel specification

Property	Method	Unit	EN590 standard	Euro 5 Diesel	EN590 B10	HVO
Cetane number	EN ISO 5165 ASTM 6890		Min 51	53	53	79
Density	EN ISO 12185	kg/m^3	Min 820	836	839	779
Aromatics	EN 12916	% m/m		20.5	15.2	0.2
Poly aromatics			Max 8	3.8	2.7	<0.1
Initial boiling point (IBP)		°C		208	170	160
Final boiling point (FBP)		°C		361	355	299
Sulfur	ASTM D-5453	mg/kg	Max 10	<3.0	6.0	<1.0
Flash point	EN ISO 2719	°C	Min 55	89	66	81
Water	EN ISO 12937	mg/kg	Max 200	36	48	60
Oxidation stability	EN 15751	hours	Min 20	51.5	>20	–
Oxygen content	ASTM D5291	% m/m		0.7	1.1	–
Viscosity at 40 °C	EN ISO 3104	mm^2/s	Min 2.0, Max 4.5	2.9	2.6	3.0
FAME	EN14078	% v/v	Max 7.0	5.2	10.2	–
Lubricity at 60 °C	EN ISO 12156	μm	Max 460	182	<400	245
CFPP	EN166	°C	Climate dependent	−19	−20	−19 (S) −35 (W)
Cloud point			Climate dependent		−10	19 (S) −36 (W)
Net heating value	SS 155138 ASTM D4809	MJ/kg		42.9	42.4	44.1

(*S*) Summer quality
(*W*) Winter quality

3 Results and Discussion

3.1 Cold Start Operability

The cold start tests were performed to investigate HVO ability to start down to −27 °C. The starting time was compared to EN590 B10 fuel. In Fig. 1 the results show that down to −15 °C there is no difference between the fuels. However below −15 °C the EN590 B10 could not start with the actual engine, but with HVO the Engine could start down to −27 °C. The explanation can be the difference in cloud point and CFPP for the fuels, since higher cloud point and CFPP causes filter blocking.

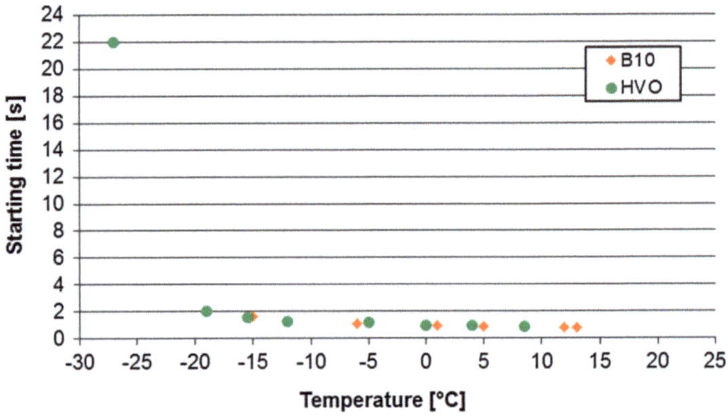

Fig. 1 Cold start for MK1 B5 and HVO fuels

3.2 Durability

A Volvo car of model V60 was used as a driving school vehicle and operated on HVO fuel for 100,000 km. The vehicle was mostly operating in city driving conditions. After the test the engine with oxidation catalyst, particulate filter, fuel system and fuel tank were disassembled and each part was analyzed in order to understand the long term effect on HVO. Comments and negative impacts for each vehicle part are shown in Table 2. Conclusion is that neither of the analyzed parts showed any negative impact.

The diesel particulate filter regenerates either passively or actively. Passive regeneration occurs when the exhaust temperature is high enough for the particulates to oxidize in the filter, typically at highway driving. At city driving however the regeneration is active which means that fuel is injected late in the expansion stroke. Some of this fuel can pass the piston rings and dilute the engine oil. If the fuel has a high distillation range, less will evaporate from the oil thus causing more dilution. Therefore city driving can be considered as worst case in terms of oil dilution.

The fuel dilution was analyzed at each service of the driving school vehicle and is plotted in Fig. 2. The level of dilution is above recommended 2 % however representative for city driving and it was concluded that HVO fuel does not cause increased fuel dilution.

3.3 Engine-Out Emission Tests

Engine test bed tests were conducted with HVO and compared with EN590 fuel with 10 % FAME (B10) at various speeds and loads. In comparison with B10 the engine-out soot emissions achieved maximum 64 % reduction which is shown in

Table 2 Long term impact on different vehicle parts

Vehicle part	Comments	Negative impact (yes/no)
Diesel oxidation catalyst	No face plugging	No
Diesel particulate filter	No soot on outlet side	No
Fuel filter	No effect of fuel ageing, no deformation on filter medium	No
Fuel tank	Clean "as new"	No
Fuel pump (low pressure)	Clean "as new"	No
Turbo system	Compressor outlet clean and little more soot than "normal" on the turbine side which indicates kind driving	No
EGR valve	Much soot but not more than normal after this driving distance	No
Cylinder head	Soot and coke layers evenly distributed	No
Cylinder liner		No
Pistons		No
Valves	Thin soot deposits on plate and shaft, but not more than normal	No

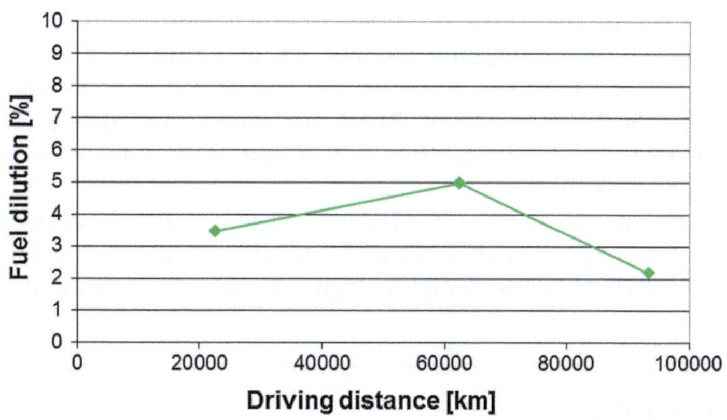

Fig. 2 Fuel dilution in engine oil

Fig. 3. The explanation is that the fuel composition has a significant impact on soot formation and flame temperature. The soot formation in flame increases with the total aromatic content, particularly with the poly-aromatic content, and since HVO does not contain any poly-aromatics it causes low soot emissions [2, 3]. Most diesel vehicles use particulate filter and if less soot is accumulated in the filter (as for low aromatic fuels) regeneration will be less frequent fuel consumption reduced.

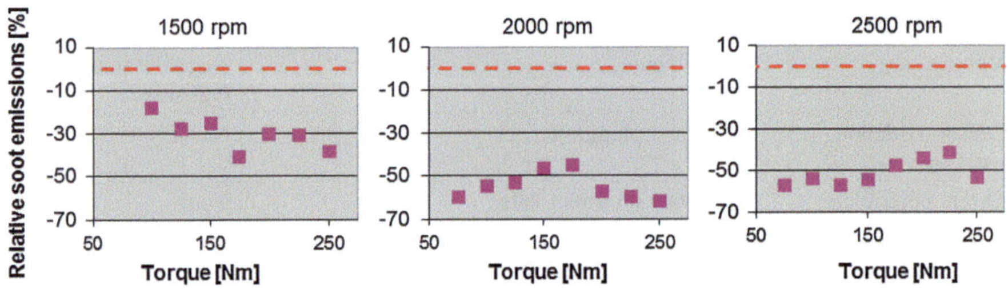

Fig. 3 Engine out soot emissions

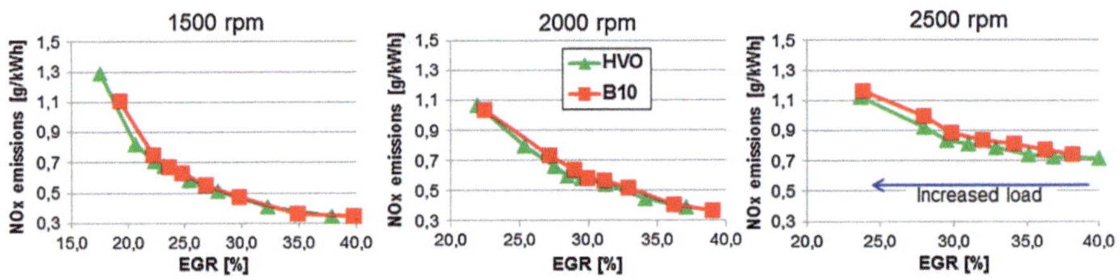

Fig. 4 NOx emissions versus the EGR

As for NO_x emissions the flame temperature has large effect on the formation. Fuels containing aromatics with high carbon to hydrogen ratio have higher flame temperature than paraffinic fuels [4]. When EGR is applied however the effect of fuel properties on flame temperature becomes less important [5]. In Fig. 4 the NO_x emissions versus the EGR are shown for three speeds and various loads. For an increased load the EGR is reduced. The figure shows small differences in NO_x emissions at 1500 and 2000 rpm while at 2500 rpm HVO reduces the NO_x up to 7 % in comparison with B10.

The specific fuel consumption for HVO in relation to B10 is shown in Fig. 5. The difference corresponds well to the difference in heating value (4 %). The high heating value of HVO fuel is caused by high H/C ratio and absence of oxygen.

3.4 Vehicle Emission Tests

In the vehicle emission tests HVO fuel was compared to Euro 5 diesel fuel. The vehicle was operated in the NEDC cycle and tailpipe emissions of HC, CO, CO_2, NOx and PM were measured as well as fuel consumption.

The HC and CO emissions were reduced by 65 and 80 % respectively for HVO fuel (see Fig. 6) which can be explained by the absence of aromatics and the high hydrogen to carbon ratio. The high cetane number of the HVO fuel could also be one reason for lower HC and CO emissions since higher cetane number reduces the

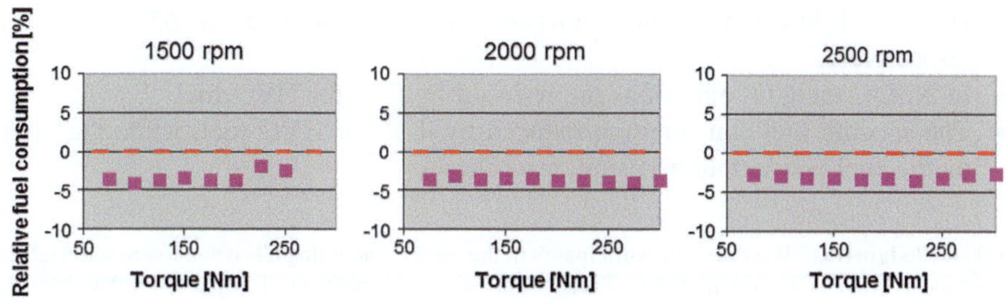

Fig. 5 Relative fuel consumption for HVO in comparison with B10

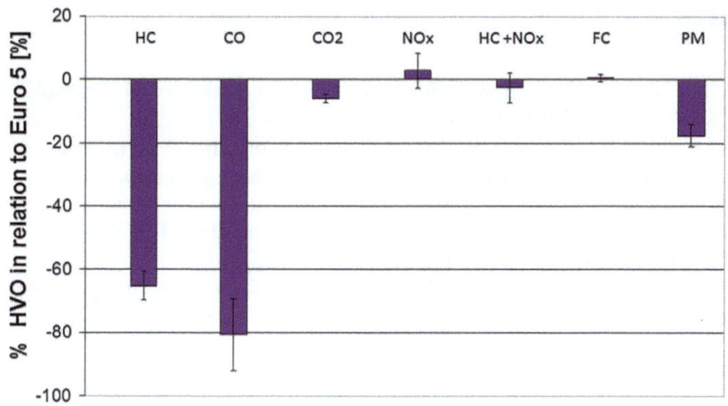

Fig. 6 Emissions from vehicle tests

ignition delay and thereby overleaning. However in this study pilot injections were used causing less effect of the cetane number. The CO_2 emissions are 6 % lower for HVO fuel due to the higher hydrogen to carbon ratio.

4 Conclusions

The conclusions to be drawn from this study, when vehicles and test bed engines were operating on HVO fuel, are the following:

- No cold start issue could be observed
- No impact after 100,000 km driving on HVO could be detected on the engine, aftertreatment system or fuel tank.
- HVO fuel do not cause extremely high fuel dilution in engine oil because of the relative low final boiling point.
- HVO fuel lowered the engine-out emissions of soot with 60 %

- HVO fuel lowered tailpipe emissions of HC and CO by 65 and 80 % respectively.
- In NEDC the CO_2 emissions are reduced by 6 % for HVO fuel
- The specific fuel consumption reduced by 4 % for HVO fuel, while the volumetric fuel consumption is similar to Diesel.

Acknowledgments We thank all participants in the project including Christina Svensson, Gerben Doornbos, Magnus Karlsson, Johan Zackrisson, Marco Gustafsson, Tuukka Hartikka, Markku Honkanen, and finally Energimyndigheten for financial support.

References

1. Hartikka T et al (2012) Technical performance of HVO (hydrotreated vegetable oil) in diesel engines. SAE paper no 2012-01-1585
2. Sugiyama K et al (2011) Effects of hydrotreated vegetable oil (HVO) as renewable diesel fuel on combustion and exhaust emissions in diesel engine. SAE paper no 2011-01-1954
3. Azetsu A et al (2003) Effects of aromatic components in fuel on flame temperature and soot formation in intermittent spray combustion. SAE paper no 2003-01-1913
4. Worldwide fuel charter, fifth edn, 2013
5. Beatrice et al C (2010) Impact of RME and GTL fuel on combustion and emissions of a "torque-controlled" diesel automotive engines. SAE paper no 2010-01-1477

Assessment of the Viability of Vegetable Oil Fuels: Species, Land, Social, Environmental, Population and Safety Considerations

I.F. Thomas, N.A. Porter and P. Lappas

Abstract This work follows the literature review and on-road trial findings presented at ICSAT 2013 (Thomas et al. Assessment of the viability of vegetable oil based fuels. In: Lecture notes in mobility—sustainable automotive technologies 2013, 2013). In addition, it addresses the availability of non-food oil-producing species and land, the sociological and environmental aspects associated with growing biofuels and the potential increased safety of using vegetable oil fuels and blends. The author claims that it is essentially impossible to provide the world's current energy needs without fossil fuels unless humanity is prepared to control its population to a sustainable level. Production methods and costings are not addressed as intended and will be the subject of further study. Costs of growing, extracting, purifying and transporting vegetable oil fuel are lower than any alternatives given the simplicity of processes involved, the fact that almost anyone can perform them and the fact that the fuel can be grown and processed close to point of use by small, non-monopolistic enterprises. Readers wishing to consider costings are referred to reports by the EC Joint Research Commission and the International Energy Agency (IEA) (Production costs of alternative transportation fuels—influence of crude oil price and technology maturity; International Energy Agency; Jensen, Unmodified vegetable oil as an automotive fuel; the institute for prospective technological studies of the European commission directorate general joint research centre).

I.F. Thomas (✉)
I F Thomas & Associates, 123 Nelson Place, Williamstown 3016, Australia
e-mail: ifta@ifta.com.au

N.A. Porter
Discipline Head Environmental Science, School of Applied Sciences, RMIT University, Swanston Street, Melbourne 3016, Australia
e-mail: nichola.porter@rmit.edu.au

P. Lappas
Lecturer Renewable Energy, School of Aerospace, Mechanical and Manufacturing Engineering, RMIT University, Plenty Road, Bundoora East 3083, Australia
e-mail: petros.lappas@rmit.edu.au

I. Denbratt et al. (eds.), *Sustainable Automotive Technologies 2014*,
Lecture Notes in Mobility, DOI 10.1007/978-3-319-17999-5_19

Keywords Straight vegetable oil (SVO) · Pure plant oil (PPO) · High-flashpoint fuel · Oil-producing species · Land availability · Land-grabbing · Population · Explosion · Fire

1 Introduction

The compression ignition engine (oil engine) was invented by Herbert Akroyd Stuart and patented in 1886 [1, 2]. The oil engine was very successful and 32,417 were manufactured (Figs. 1 and 2) [4–6]. This was before Rudolf Diesel developed and patented his improved, higher pressure design. An engine designed specifically for running on vegetable oil was developed by Ludwig Elsbett in 1977 (Figs. 3 and 4) [7]. It contains three cylinders, a two-part piston, bowl-in-piston swirl chamber and a novel method of cooling. Production ceased following progressive re-application of fuel tax in Germany from 2008 to 2012.

Now that governments are beginning to realise the importance of sustainability however, vegetable oil fuel is again being considered.

Fig. 1 Hornsby-Akroyd vapourising oil engine: Richardhornsbyvaporizingoilengine.jpg

Fig. 2 Herbert Akroyd Stuart: Tractor and Construction Plant Wiki. http://tractors.wikia.com/wiki/Herbert_Akroyd_Stuart

Fig. 3 The Elsbett Motor. http://de.wikipedia.org/wiki/Elsbett-Motor

Fig. 4 Ludwig Elsbett. http://en.wikipedia.org/wiki/Ludwig_Elsbett

2 Literature Review

139 papers by authors from 32 countries have been reviewed covering the period 1980–2014 (Table 5). Exhaust emissions for vegetable oil are improved in 91 instances and are worse than diesel fuel in 54 instances (Table 1). 47 instances variously show improved brake power, brake thermal efficiency, brake specific fuel consumption or torque compared with diesel fuel and worsening in 16 instances (Table 2). Several researchers describe adverse impacts of the use of vegetable oil fuels on the engine and its componentry caused by high viscosity, presence of free fatty acids and propensity to polymerise, particularly where 100 % vegetable oil is used. They suspect that this will prejudice use of vegetable oil particularly in the long term. Saturated oils are less likely to polymerise but tend also to be higher in viscosity and more likely to be solid at room temperature. Chemical properties affecting cetane number, presence of hazardous emissions and late combustion enhancement are the aromatic/aliphatic compound balance and the amount of combined oxygen present [8]. The Czech Republic has operated fleets of dual-fuelled vehicles successfully since 1977 [9]. Few long-term trials have been conducted in other countries. On balance there appears to be wisdom in using vegetable oil either alone or in blends with more volatile fuels for both automotive and stationary engine applications.

Table 1 Emission and noise findings; red (U) up, green (D) down, white (S) same

CO	CO_2	HC	PM	NOx	Smoke	Noise	Oil type etc.
U	–	U	U	–	–	–	Sunflower
D	–		S	U	–	–	Shale oil
D	D	–	–	U	–	–	Jatropha
S	S	S	S	D	S	–	Karanja
D	–	D	D	–	–	–	Diesel/water emulsion
D	–	D	–	–	D		Rapeseed
–	–	–	–	D	–	–	Karanja
U	–	U	–	D	D	–	Karanja, Ricebran, Neem, prod gas
U	U	U	–	D	–	–	Jatropha
D	–	–	–	U	D	–	Ethanol, propanol
U	–	U	–	–	U	–	Hazelnut
–	–	–	–	D	–	D	Cotton, linseed, peanut
S	S	–	–	S	–	–	DFO, cottonseed
U	–	–	–	S	–	–	Karanja
U	D	U	–	D	U	–	Poon
U	–	D	D	U	–	–	Rape, sunflower, false flax, jatropha
S	S	S	S	S	–	–	10 % cottonseed oil
S	S	S	D	S	–	–	Cotton, rape, sunflower
D	S	D	–	U	D	–	Jatropha
D	–	–	D	D	D	–	Putranjiva
D	S	D	D	D	D	–	Putranjiva, Karanja
D	–	–	–	D	D	–	Rape heated to 100 °C
S	–	S	–	–	D	–	Diesel, ethanol, butanol
D	–	D	–	U	D	–	Waste corn oil
–	–	U	–	D	U	–	Turpentine oil
U	–	U	–	D	U	–	70/30 CNSO/CMPRO
S		D	–	D	U	D	RO, lard, chicken fat
–	–	–	–	–	D	D	Linseed, alcohols
U	–	U	U	D	–	–	Canola
–	–	–	–	D	D	–	Coconut
U	–	U	–	D	–	–	LPG, DEE
S	–	S	–	D	S	–	Soapnut best at 10 %
–	–	–	–	D	–	–	Rape, palm, animal fat
–	–	–	–	D	–	–	Saturated VOs
U	D	–	–	–	D	–	Heated sunflower
D	–	–	–	–	D	–	WVO
D	–	D	–	U	D	–	Orangeskin powder
U		D		U	D	–	100 % Orange oil
U		D		D	D	–	36 mg/s DEE + Orange
U	–	D	–	D	U	–	100 % cottonseed
CO	CO_2	HC	PM	NOx	Smoke	Noise	Oil type etc.
U	–	U	–	D	D	–	80/20 cotton/butanol and cotton/DEE
U	–	U	–	U	D	–	Sun, cotton, corn, olive
D	–	D	–	D	–	–	For max 5 % canola oil
–	D	–	–	–	–	–	Rapeseed
D	–	D	–	U	–	–	Soy, Rape, RME
U	–	U	U	D	–	–	Traffic; canola
D	–	D	D	D	–	–	Country; canola
–	–	–	–	–	D	–	Soybean
U	–	D		S	–	–	Peanut, canola and sunflower oils
U	–	D		U	–	–	
U	–	U	–	S	D	–	Diesel, DME
–	–	–	–	–	D	–	Rapeseed

Table 2 Sample of the performance summary

Fuel	Purpose	Engine	Cyl	Comp ratio	Inj press bar	Engine mods	Performance versus diesel
Jatropha	Auto	DI	–	–	–	Preheat	Viable
Karanja	Agric	Kirloskar	1	–	180–200		Better up to 50 % oil
Pistacia	–	Lister 8/1, 6 kW, IDI	1	17.5:1	–	–	BTE, brake power down; BSFC up
Rapeseed 100 %	Boiler	Roca AR/ 25GT	–	–	–	–	Viable
Sapodilla and Teitai tree oils	–	–	–	–	–	–	Viable @ 10 % oil; of high calorific value
Sunflower and cottonseed/ diesel	Auto	Rainbow 186	1	18:1	196	ZrO_2 coated	All fuels viable; SFO better than CSO
Karanja	Auto	6 kW, 4-stroke	1	–	–	–	Viable @ 10 % oil; BTE up; BSEC down
Karanja oil, producer gas	All	Apex Kirloskar	1	17.5:1	200–225	–	Brake thermal efficiency (BTE) up
100 % WVO run for 500 h	Agric power	Listeroid c/w generator	–	6:1	–	75 % ld 650 rpm	500 h trial; Oil-change every 110 h

3 Availability and Status of Non-food Oil-Producing Species

Vegetable oil fuel is presently being sourced entirely from food oils, principally rapeseed (Canola) oil. Unless this changes, it will ultimately lead to mass starvation. This is neither desirable nor necessary. A listing of 204 non-food oil-producing species and 73 species cultivated principally for food production has been compiled and is available upon request. The distinction between these two sub-sets is not rigid as all species have some use and often that includes use as food or food supplementation. Of the non-food species 18 are salt tolerant and 34 are capable of growing in arid areas (Tables 3 and 4). Of the total of 277 species 175 are trees, bushes or large shrubs and 102 are ground crops. The latter are easier and therefore cheaper to harvest but tree crops once established produce for many years without the need to re-sow. Ground crops are mostly grown as annual plants. Thirty of the non-food species have an average seed yield of 0.95 t/ha/year and 27 food-related species yield an average of 1.38 t/ha/year. These are sufficiently similar to suggest that with cultivation, there is oil available for fuel use without using food oils. Highest yielding non-food species are Torchwood (*Canarium indicum*, 4–7 t/ha/

Table 3 Salt tolerant species

	Common name	ICN name
1	Arak or Toothbrush tree	*Salvadora persica*
2	Beaded Samphire	*Sarcocornia quinqeflora*
3	Cabbage Palm	*Sabal palmetto*
4	Glasswort	*Salicornia persica*
5	Long-spiked Glasswort	*Salicornia dolichostachya*
6	Marsh Samphire	*Salicornia bigelovii*
7	Perennial Saltwort	*Sarcocornia fructicosa*
8	Pinot peanut	*Arachis pintoi*
9	Saltmarsh Mallow	*Kosteletzkya virginica*
10	Sand Rocket	*Diplotaxis tenuifolia*
11	Scurvy Grass	*Cochlearia officinalis*
12	Sea Asparagus	*Salicornia brachiata*
13	Sea Buckthorn (Common)	*Hippophae rhamnoides*
14	Sea Mango tree	*Cerbera manghas*
15	Sea or Bengal Almond	*Terminalia catappa*
16	Sea Radish	*Raphanus raphanistrum*
17	Sea Rocket (American)	*Cakile edentula*
18	Sea Rocket (European)	*Cakile maritima*

year) [10] and Pongam, Karanji, Honge or Indian Beech tree (*Milletia pinnata*, 3–5 t/ha/year) [11]. Highest seed yielding food-related species are the Oil Palm tree (*Elaeis guineensis*, 5.95 t/ha/year) and Coconut Palm tree (*Cocos nucifera*, 2.67 t/ha/year).

Of the thirty non-food species, the average seed oil content is 34.8 % and for three food-related species, the average is 47.7 %. Highest non-food seed oil content species are Kukui Nut tree (*Aleurites montana*), 59 % [12], Bagilumbang tree (*Aleurites trisperma*), 55 % [13], Hunters Nuts tree (*Omphalea megacarpa*), 57 %, Torchwood (*Canarium indicum*), 50.9 % and Mahwa tree (*Madhuca indica*), 50 %. The highest food-related species seed oil content is Oil Palm (*Elaeis guineensis/E. oleifera*), 49 %. These figures also indicate the potential viability of using non-food species for fuel.

It is the combination of salt tolerance and drought resistance of some of the plants which leads them to be particularly promising for fuel production in desert/semi-desert areas and also in arid and semi-arid coastal margins. Plants of the Salicornia genus are being heavily researched and several trials are underway. Some examples where cultivation of biocrops has been performed or planned are Salicornia spp. and *Salltmarsh Mallow* (*Kosteletzkya virginica*) in Bahia Kino, Sonora, where planned production is 2600 t/year for biodiesel and 52,500 t/year for solid biofuel from 480 ha. Genetic research is also underway here to improve cetane number, oxidative stability, cold-flow properties, lubricity and productivity. An integrated seawater village is planned combining living, recreation, tourism and

Table 4 Arid area growing species

	Common name	ICN name
1	Arak tree	*Salvadora persica*
2	Argan tree	*Argania spinosa*
3	Beaded Samphire	*Sarcocornia quinqueflora*
4	Ben tree	*Moringa peregrina*
5	Bladderpod	*Physaria fendlerii*
6	Blue Flax	*Linum lewisii*
7	Brahea Palm	*Brahea salvadorensis*
8	Buffalo Gourd	*Cucurbita foetidissima*
9	Artichoke Thistle	*Cynara cardunculus*
10	Crown flower	*Calotropis gigantea*
11	Desert False Indigo	*Amorpha fruiticosa*
12	Desert Sunflower	*Helianthus anomalus*
13	Egusi	*Cucumeropsis mannii*
14	Firestick/Pencil tree	*Euphorbia tiraculli*
15	Ghaf tree	*Prosopis spicigera*
16	Glasswort	*Salicornia persica*
17	Horseradish tree	*Moringa oleifera*
18	Jojoba	*Simmondsia chinensis*
19	Karir tree	*Capparis decidua*
20	Marsh Samphire	*Salicornia bigelovii*
21	Milkweed	*Asclepias syriaca*
22	Mongongo Nut tree	*Schinziophyton rautanenii*
23	Mustard Greens	*Brassica juncea*
24	Palestine Pistachio	*Pistacia palaestina*
25	Patawa tree	*Jessenia bataua*
26	Pinot Peanut	*Arachis pintoi*
27	Quechua or Goosefoot	*Chenopodium quinoa*
28	Royle tree	*Prinsepia utilis*
29	Sea Buckthorn tree	*Hippophae rhamnoides*
30	Sea Radish	*Raphanus raphanistrum*
31	Sea Rocket, American	*Cakile edentula*
32	Sea Rocket, European	*Cakile maritima*
33	Sesame	*Sesamum indicum*
34	Sprengel Seed	*Euphorbia lagascae*

farming of shrimps, seaweed, bivalves, fish, *Salicornia*, mangroves, microalgae, Artemia and salt [14–16].

Some other cultivation activities are *Salicornia*—4500 acres in Eritrea, also in Ras al-Zawr Saudi Arabia, Kahuku Hawaii, San Francisco Bay California, Dawlish Warren UK, Texel Holland, Wilhelmshaven Germany and Yuma, USA; *Jatropha* —80,000 acres in India, 75,000 acres in Brazil, also in Costa Rica, Dakatcha

Table 5 Papers reviewed by country

Country	Papers
USA	21
India	24
Turkey	12
Greece	7
Germany	7
France	6
Spain	6
China	6
Brazil	4
Burkina Faso	4
UK	4
Jordan	3
Romania	3
Nigeria	2
Australia	2
Italy	2
Czech Republic	2
Canada	1
Denmark	1
Egypt	1
Finland	1
Ghana	1
Indonesia	1
Japan	1
Lithuania	1
Netherlands	1
South Africa	1
South Korea	1
Slovak Republic	1
Switzerland	1
Ukraine	1
Yugoslavia (former)	1
	139

Woodland Kenya, Tamil Nadu, Ghana and Uganda. Other species being experimentally grown are 250 acres of *algae* in Karratha Australia *microalgae* in India, *Dunaliella microalgae* in Israel, *Miscanthus* in Iowa USA; *Sweet Sorghum* in Brazil; *Switchgrass* in Loudon County USA, *Camelina* in USA, *Sask Mustard* in Canada, *Giant Reed* in USA and *Sarcocornia* in Alicante Spain.

4 Land Available for Renewable Energy Production

A major concern associated with the production of biofuels is the current reliance on rapeseed, sunflower, cottonseed and palm oils. There is competition between food and fuel uses of these edible oils, posing somewhat of a moral dilemma. The issue is being debated in scientific journals as well as in the popular press but there are few actual scientific studies that help to clarify this [17]. Because food crops have been cultivated for many years, it may actually be wiser to use these for fuel by producing more of them, than by opting to use new, less understood non-food species. That is, apart from their need for arable land.

The global land surface of 256 countries is 14,900 Mha [18–20]. Human related land-use is between 7700 and 9900 Mha. Of this, 5000–7200 Mha is unused wilderness of which 3200 Mha are declared protected areas and are unavailable leaving 1800–4000 Mha of wilderness available. Currently unused, potentially productive land which may be considered to be available for the cultivation of sustainable energy crops is 400 Mha [17]. In energy terms, the current world requirement is 500 EJ/year. If biomass was to completely replace current fossil-derived fuels, 2500 Mha would be required and this is clearly not possible [17]. Many of the arid area growing species are declared weeds in certain countries and others which have not yet been cultivated may become invasive and prove difficult to eradicate. Cultivating previously non-cultivated lands also results in loss of soil black-carbon taking many years to recover. An example of this is the draining of peat bogs to create farmland [21]. Notwithstanding these difficulties, potential for energy production by new sources has been assessed by various workers (Hoogewijk, Smeets, Doornbosch, Faaij) [17]. Collectively, their predictions range from 30 to 1400 EJ/year with an average of approximately 400 EJ/year. In the next 21 years to 2035, world energy use is predicted to increase by 20 % from 500 to more than 600 EJ/year according to the World Nuclear Association [18]. The latter organisation advocates the use of nuclear power both because it is an extremely dense form of energy and because it saves other natural resources. WNA also describes nuclear energy as 'clean'—correct in CO_2 emission terms but the author suggests, not in any other sense. The author quotes his Engineers Australia Risk Engineering Society colleague, 'If the Romans had had nuclear power, we would still be dealing with their waste'. The author considers the use of nuclear power to be inappropriate until a safe process is found which also does not produce radioactive waste.

By taking the average future energy value of 400 EJ/year postulated above, it is possible for the world to create sufficient energy to last until 2035. WBGU [17] states that 34 % of world land area is used for agriculture (5000 Mha). Of this 1500 Mha is arable, 2500–3500 Mha is pasture, 20 Mha are used for growing fuel, and cities occupy 200–300 Mha, usually located on low-lying prime agricultural land (Faaij, Salvatore) [17]. 80 % of farmland is used for rearing cattle which provide only 17 % of world food needs (Steinfeld) [17]. In the 40 years from 1963 to 2003, farmland has increased by 460 Mha.

If we ignore the politics and complexity of energy distribution, use of a 600 km × 600 km area of the Sahara Desert (40 Mha) could create current world energy needs using photovoltaic solar power generation [17]. A similar area would be required using solar thermal energy production [22]. This in area terms alone appears possible. A 20-member 2009 German-led international consortium is planning the Desertec industrial initiative formed in Germany as Dii GmbH, to develop concentrated solar power (CSP) in the Sahara Desert. This is a combination of mirror arrays, heat-transfer fluid filled receivers and steam-turbine power generation using natural gas at night [23]. The plan is to distribute power to Middle East and North African (MENA) countries as well as Europe and Iceland using high voltage direct current transmission. It is understood that the Moroccan first stage is being constructed. The consortium claims that 0.3 % of the Sahara Desert (2.8 Mha) would be sufficient to power Europe. Political and energy transmission realities and energy storage realities are likely to mean that solar power is not the complete answer. Rather, where arid land or land-sea margins can be used to grow oil-producing species, this should be encouraged provided that adequate steps are taken to prevent social upheaval, escalating commodity prices and reduced biodiversity. Equally where countries have a high incidence of solar radiation (e.g. Australia), its use should be encouraged. A reportedly more efficient way of producing biofuel is by the use of algae for example in Brazil where 5000 L/year can be manufactured from a reactor containing 3.5 km of tubes occupying 10 m^2 of land [24]. Similarly the race is on to find decomposer fungal enzymes which can help to produce second generation biofuels more efficiently [25] and to produce biofuel from seaweeds.

5 Social and Environmental Considerations

Countries which have a mandate to produce biofuel to offset the rising cost and reducing availability of fossil fuels following the 1992 EU-US Blair House Agreement [26] and limited ability or desire to use their own lands for fuel production, are buying land in other countries on a large scale. This 'land grabbing' is unregulated and less than transparent. The International Land Coalition (ILC) is drawing attention to this phenomenon. ILC's vision is that 'secure and equitable access to and control over land reduces poverty and contributes to identity, dignity and inclusion' [27]. Europe is described as the central driver of this land acquisition because it imports much of the raw materials it uses. In addition some 27 countries as well as the EU, Brazil and the US have biofuel mandates. Eleven countries mainly in East Africa and Southeast Asia account for 70 % of the 'grabbed' land according to ILC. From 1006 transactions in the period 2000–2012, ILC shows that 70.2 Mha of land were acquired as claimed, equal to half of the size of Western Europe. China announced plans in 2010, to set up ethanol projects in a range of African countries. Malaysia's *Sime Darby* and Finland's Singapore-based *Nesté Oil* are claimed to be exploiting palm oil to meet European needs according to Grain [28]. Sime Darby has developed 500,000 ha in Malaysia for palm oil production and plans a further 220,000 ha in Liberia. In the first 10,000 ha stage of this development 15,000 people were displaced [28]. In Gabon, Singapore-based 'Olam' plans to clear 50,000 ha for a palm oil plantation within a 300,000 ha concession area provided by the Gabon government. In areas like this as well as deforestation and displacing of people, available water is being consumed by the plantations and wildlife species such as gorillas and orangutans are under threat. Some other examples of countries where peoples are being displaced are Kenya, Sierra Leone, Guinea, Brazil, West Papua and Colombia.

Large-scale cultivation of species which have been used for thousands of years by local populations can adversely affect them in other ways as well as displacement. Their normal sources of herbs, food and medicine can become unaffordable resulting for example in a drop in per capita caloric intake. To ensure minimisation of these adverse social impacts WBGU [17] has proposed a series of 'guard rails' for example recommending that 'the amount of land available must be sufficient to ensure a daily intake of all people, of 2700 kcal'. Given that over 900 million people are presently 'food insecure' this means that all available farmland must be used primarily to produce food. 'There are alternative sources of energy but there are no alternatives for food' [17]. A second guard rail is proposed to ensure that 'all humans receive between 700 and 1000 kWh annually to meet their basic energy needs'. A third guard rail proposal addresses health and its continued good state as being 'a fundamental human right'. Without enough food in the world presently it is difficult to see how we can also provide fuel, even with our ability to continually improve production efficiencies.

6 Population

Human population has been growing exponentially over nearly all of recorded time. This cannot continue even if crop yields were also to grow exponentially. Therefore the solution to the current food plight, overcrowding, competition for resources, the fuel dilemma, impacts on climate, impacts on biodiversity and species extinctions, is for our population to stabilise. Not only must this happen, but it must also stabilise at a level which is sustainable. The difficulty in controlling world population is the fact that we live in many autonomous regimes. WBGU in its 'Summary for policy makers', advocates the setting up of a 'Global commission for sustainable land use'. Clearly a wise initiative but one which the author considers should be preceded by the setting up of a 'Global commission for a sustainable population'. No country will be inclined to stabilise or reduce its population in isolation, given fears that other countries would take advantage. The effort must be universal.

Sustainable Population Australia Inc recently stated in its February 2014 newsletter "Most problems in the world today, such as climate change, stem from one immense problem which seems to be the 'elephant in the room' that no-one wants to talk about. This problem is our ever expanding population" [29]. Current world population occupying 244 countries is 7.13 billion people [30, 31]. Human beings were described recently by the famous environmentalist Dr David Suzuki as a 'cancer on the planet' [32] and similarly by Paul Ehrlich in the 1970s in his book 'The Population Bomb' [33]. The population is predicted to increase by 22 % from 1999 to 2030 (UN Food and Agriculture Organisation) [17]. 923 million people are presently 'food insecure' [17, 34] and China with 1300 million people will more than double its energy use by 2030. The rate of world population growth and the extent to which it is accelerating is indicated by the time taken to double the population. The doubling time in 1715 from 0.38 to 0.75 billion was 544 years, in 1804 it was 304 years, in 1881 166 years, in 1927 123 years, in 1960 79 years, in 1974 47 years and in 1999 the population doubling time from 3 to 6 billion was 39 years [35].

7 Potential for Improved Safety by Using High Flashpoint Fuels

Vegetable oil flashpoints range from 150 °C (Babassu oil) to 400 °C (Cottonseed oil). These are higher than any other kind of fuel for example biodiesel 130 °C, diesel 62 °C, ethanol 16.5 °C, methanol 11 °C, dimethyl ether −41 °C, petrol −43 °C, diethyl ether −45 °C, LPG −104 °C, CNG/LNG −188 °C and hydrogen −253 °C. All current fuels are more dangerous than any of the vegetable oils. Diluting vegetable oil with diesel fuel reduces the flashpoint for example to between 110 and 120 °C for a 50/50 blend [36]. The impetus in terms of safety should therefore be to use pure, undiluted plant oil.

An attempt is made to establish the reduction in loss of life which might occur if all fuels were vegetable oil by assessing the number of deaths caused by events involving current fuels and fuel-like substances. Two sources of data are used namely road accidents involving fire [37–41] and 616 non-natural accidental explosions which occurred in the period 1910–2013 (Fig. 5) [39, 42–53]. The full explosion list is available on request.

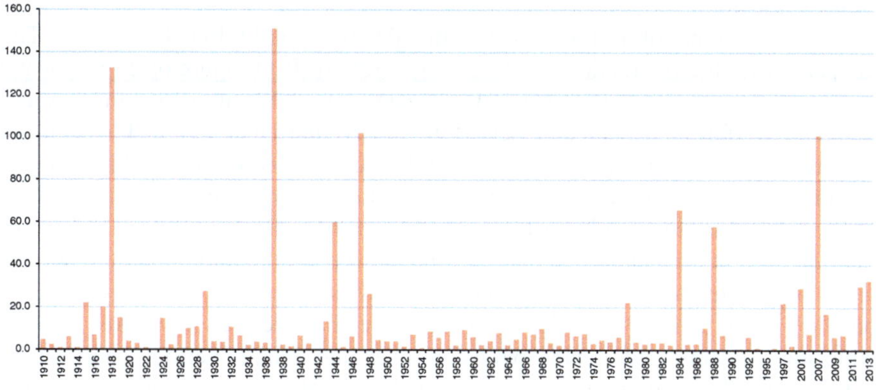

Fig. 5 Deaths per explosion for 616 events 1910–2013, which caused 11,466 deaths

Analysis of these events shows that over time, explosions keep happening with high and low loss events occurring in a cyclic manner. Generally, the greater the confinement the greater is the extent of harm caused, with the exception that some of the most devastating explosions occur in the open air as so-called unconfined vapour cloud explosions (UVCEs). With munitions designed to be harmful, the confinement is of the explosive material in its casing. With mines, tunnels and any form of building, the confinement is by the enclosing structure. In the open air, the change from deflagration (burning) of a cloud of vapour to detonation (explosion) depends upon the extent to which obstacles are encountered as the flame-front progresses. The change from deflagration to detonation occurs when the flame velocity reaches and then exceeds the speed of sound [54–57].

Natural gas caused most of the explosions followed by petrol, propane, butane, solvents, dust, LPG, kerosene, ethylene, acetylene, hydrocarbons and crude oil. Natural gas also caused the most deaths followed by agricultural chemicals, LPG, petrol, propane, dust, solvents, LNG, propylene and diethyl ether. The greatest number of deaths per event was caused by diethyl ether, followed by agricultural chemicals, LNG, propylene and LPG. Although factories have the most events and cause the most deaths, the worst events in terms of deaths per explosion are those involving pipelines, roads, and large scale occupied premises. To prevent this, it is reasonable to suggest that reticulation of gases by pipeline and gas supply to multiple occupancy premises should be reconsidered. The US has the highest number of events and of deaths. This does not suggest that this country is less responsible rather it shows that its data is more readily available internationally. In contrast, the only event listed for North Korea is one which remained unreported for a considerable period and where help from other countries was initially declined. The event however appears to have been so large that the country acceded to allowing access to the International Red Cross. It was not until much later that North Korea published details of what had happened. Deaths per explosion were highest for the former USSR followed by Congo, North Korea, Mexico, Germany, Bangladesh, Ukraine, Venezuela, Spain, Kenya, Iran, Singapore, Turkey, Nigeria, Thailand, China and South Korea. Twenty four other countries registered 50 or less deaths/explosion.

The total number of deaths caused by the 616 events is 11,466 (111.3/year). This compares with the International Labour Organisation (ILO) figure of 22,898 deaths in 25 years from 1969 to 1993 (915.9/year) caused by all non-natural causes including warfare and malicious acts [50]. Clearly the latter figure should not be used in the present context. However if the ILO analysis is correct, the higher figure provides a glimpse into the totality of deaths which might be caused per year from all fire-related causes involving fuels. Neither of these figures will allow for non-reporting and the reluctance of many countries to publish at all. Openly reporting, analysing and learning from accidents when they happen is the only way to ensure that they do not recur or at least, to minimise that recurrence. Charts showing explosions, deaths and deaths/explosion over the whole period, analysed by country, by substance involved and by premises involved are available on request.

Current world population is 7.2 billion people [58]. Using the UK and US derived figures of 0.71 and 0.94 road accident fire-caused deaths per million population and assuming that the rest of the world is the same as the UK and the US yields a potential loss-of-life in the range 5112–6768 persons/year. If all fuel was replaced with vegetable oil, the potential for loss-of-life saving is all of these plus 111.3 persons/year from explosions established above. This is a loss-of-life saving in the range 5223–6879 persons per year. The reality presently because biofuel as a whole is such a small portion of total energy, is that the non-volatile fuel use advantage will be diminished to an estimated 0.5 % × 5223–6879 = 26.1–34.4 persons/year. This assumes that all biofuel is straight vegetable oil whereas most in fact is biodiesel and blends of biodiesel with conventional fuels.

8 Conclusions

1. The developed world has some interest in biofuels particularly in Germany where a rapeseed oil fuel standard and many service outlets are available. Sixty two countries have set objectives to use a certain percentage of sustainable fuel by defined future dates. This however appears to be causing a problem rather than solving one (as is purported to be the justification), through the impacts on people occupying lands in developing countries displaced by the need for growing developed world fuel crops.
2. Developed countries are likely to be influenced in their relative lack of enthusiasm for biofuels, by ample availability of fossil fuels. In Australia which has a wealth of conventional fuels available, development is of progressively more volatile fuels some of which are not sustainable for example mineral LPG, CNG, LNG, CTL and GTL.
3. Relatively fuel-starved developing countries are enthusiastically researching into the use of vegetable oil. This enthusiasm is not limited to any one particular use or fuel type but encompasses numerous vegetable oil species both as neat oils and in blends for use in vehicles, in agriculture and in stationary engines such as for power generation.
4. The literature review shows that exhaust emissions vary when vegetable oil is used both alone and in blends. However considerably more workers claim decreases rather than increases in emissions (91 vs. 54). Engine performance parameters are either equal to or improved over diesel fuel at least in the short term. Long term, most workers have the view that difficulties associated with the higher viscosity of vegetable oils, greater difficulty in producing conventional fuel spray patterns, acidity and polymerisation have yet to be overcome. This situation will improve only when large-scale on-road trials are conducted using engines specifically designed for vegetable oil. Work done by Kleinova in the Slovak Republic however [9], monitored 171 dual-fuel adapted conventional trucks over 2 years travelling a collective distance of 2.9 million km and found them to be without problems running on 100 % vegetable oil. Nevertheless an

engine known as the Elsbett Motor was developed in 1977 and produced for a number of years until interest in vegetable oil fuel waned. It could easily be reinstated as demand for sustainable fuel rises.

5. There are numerous non-food vegetable oil species available for consideration some of which are already being grown in trial plantations. Of the 277 species addressed, 25 have been researched in test engines and found to be viable. Australia is an ideal candidate for trial growth given its large diversity of climate zones and the fact that many of the non-food species are either indigenous to the country or capable of being grown. For Australia to do this it might need to moderate its current attitude towards so-called 'weed' species for example Artichoke Thistle [59], Sea Radish [60], Indian Beech [11], Horseradish tree [61] and Lincoln weed [62]. The author considers that the greatest potential is for salt-tolerant land-sea margin growing species such as Sea Asparagus, Common Sea Buckthorn, Sea Mango, Sea Almond, Sea Radish and Sea Rocket.
6. There is only enough land available to meet some 16 % of the demand if all energy is to be produced in the form of biofuels. The practice of 'land grabbing' in developing countries by developed nations lacking sufficient of their own agricultural land has been highlighted. It is clear that presence of people and wildlife 'in the way' of these biofuel needs, take second place. This practice needs to be regulated. The continuing exponentially rising world population and the associated inequalities of consumption between developed and developing countries have led to fundamental difficulties such as inequitable availability of food, energy and sanitation. All of the world's human-caused difficulties would be overcome if we controlled our population to a sustainable level. To achieve this, the current economic mantra needs to change from 'growth' to 'sustainability'.
7. The increased safety associated with the use of non-volatile, non-flammable fuels such as straight vegetable oil has been highlighted. Here, the findings of an appraisal of explosions involving fuel and fuel-like substances over a 103-year period together with US and UK annual vehicle-fire accident statistics are combined to postulate the potential life-saving if all fuels were non-volatile. Without suggesting that these are the ultimate figures given the limitations of the data obtained, between 5000 and 7000 lives per year would be saved if non-volatile fuels were to replace all other fuel types.
8. For this sustainable, safer fuel to become truly viable, a much more intense research effort is needed, backed by and performed by vehicle manufacturers, engine manufacturers, fuel producers and governments.
9. For the human world to commence acting in a responsible, truly sustainable manner, existing world organisations would be required to act decisively and quickly. New world organisations with the power to act may also need to be brought into being.

References

1. Thomas IF, Porter NA, Lappas P (2013) Assessment of the viability of vegetable oil based fuels. In: Lecture notes in mobility—sustainable automotive technologies 2013. Proceedings of the 5th international conference ICSAT 2013. Springer International Publishing, Switzerland
2. Production costs of alternative transportation fuels—influence of crude oil price and technology maturity; International Energy Agency
3. Jensen P (2005) Unmodified vegetable oil as an automotive fuel; the institute for prospective technological studies of the European commission directorate general joint research centre
4. Stuart HA (1864–1927). http://adb.anu.edu.au/biography/stuart-herbert-akroyd-8705
5. F.T.C.P.W. Herbert Akroyd Stuart 1864–1927. Inventor of the hot bulb heavy oil engine. http://tractors.wikia.com/wiki/Herbert_Akroyd_Stuart
6. The emergence of the oil engine. http://www.enginemuseum.org/eoe.html. Accessed 09 Apr 2014
7. Elsbett-Motor. http://de.wikipedia.org/wiki/Elsbett-Motor. Accessed 08 Feb 2014
8. Lance DL, Andersson JD (2004) Emissions performance of pure vegetable oil in two European light duty vehicles. SAE. 2004-01-1881
9. Kleinova A et al (2011) Vegetable oils and animal fats as alternative fuels for diesel engines with dual fuel operation. Fuel Process Technol 92:1980–1986
10. Thomson LAJ, Evans B (2013) Species profiles for pacific island agroforestry—Canarium indicum [cited 25 Jan 14]. http://agroforestry.net/tti/Canarium-canariumnut.pdf
11. Weed risk assessment; Milletia pinnata syn. Pongamia pinnata (Department of Employment, Economic Development and Innovation, Biosecurity Queensland) 2010
12. Aleurites montana (Lour.) Wils. http://www.hort.purdue.edu/newcrop/duke_energy/Aleurites_montana.html. Accessed 26 Jan 2014
13. Jamieson GS, McKinney RS (1935) Bagilumbang or Soft Lumbang (Aleurites trisperma) Oil
14. Mota-Urbina C (2014) GCIRO; Project Salicornia: halophyte cultivation in Sonora (USIJI Uniform Reporting Document) [cited 18 Jan 2014]. http://www.gcrio.org/usiji/pdf/Mexico-3%203-04.pdf
15. Hot energy crops for advanced biofuels: what's happening? 2014 Biofuels digest
16. Global Seawater—Alimentos, Energia, Desarrollo inmobiliario, Calentamiento global, Secuestro de CO_2, Evita la elevacion del nivel del mar
17. Schubert R et al (2011) German advisory council on global change, future bioenergy and sustainable land use (WBGU), vol 365. Earthscan, London
18. CIA World Factbook (2012) https://www.cia.gov/library/publications/the-world-factbook/index.html
19. Land area of all countries of the world [cited Dec 2013]. http://world.bymap.org/LandArea.html
20. Introduction to the hydrosphere, Chapter 8, PhysicalGeography.net
21. Cunningham M (2014) Ten disadvantages of biofuels. How stuff works. Accessed 11 Feb 2014
22. Mueller-Steinhagen H et al (2011) Methodology for optimized operation strategies of solar thermal power plants with integrated heat storage. Solar Energy 85:653–659
23. Solar energy from the sahara desert could power the world—but will it? The DESERTEC Foundation project Dii
24. This is Brazil, Episode 1, SBS Television. http://www.sbs.com.au/ondemand/. Accessed 02 May 2014
25. Magic Mushrooms, SBS Television. http://www.sbs.com.au/ondemand/. Accessed 18 May 2014
26. Blair House Agreement (1992) Glossary of Statisitical Terms, OECD
27. The land matrix: new data on large-scale land deals: not a "bubble". 2012. Landcoalition.org

28. Against the grain: land grabbing for biofuels must stop: EU biofuel policies are displacing communities and starving the planet. 2013 1–24
29. Newsletter No. 114, February 2014. Sustainable Population Australia Inc 2014 [cited Number 114; Feb 14]
30. US and World Population Clock [cited Dec 2013]. http://www.census.gov/popclock/
31. US Census Bureau website [cited 19 Jan 2014]. http://en.wikipedia.org/wiki/United_States_Census_Bureau
32. Growth is the "cancer of the planet" claims David Suzuki; Administrative Sciences Association of Canada Conference, 10–13 May 2014. http://www.muskokaregion.com/news-story/4518048-growth-is-the-cancer-of-the-planet-claims-david-suzuki/. Accessed 03 Jun 2014
33. Population 7 Billion (2011). http://ngm.nationalgeographic.com/2011/01/seven-billion/kunzig-text#. Accessed 18 Feb 2014
34. FAOSTAT new pilot version [cited 19 Jan 2014]. http://faostat3.fao.org/faostat-gateway/go/to/home/E
35. Years to double population. http://en.wikipedia.org/wiki/World_population
36. Thomas AC (2014) Flashpoint prediction using Activity Coefficients, Chemicalia Pty Ltd. Accessed 21 May 2014
37. Chowdhury N, Rowe I, Sayer G (2012) Fire statistics Great Britain, 2011–2012, 1–65
38. Fires in the U.S. (2012) http://www.nfpa.org/research/reports-and-statistics/fires-in-the-us. Accessed 30 Jan 2014
39. Multiple-death fires-deadliest fires or explosions in the world (2011) http://www.nfpa.org/research/reports-and-statistics/fires-in-the-us/multiple-death-fires/deadliest-fires-or-explosions-in-the-world. Accessed 30 Jan 2014
40. Fire Causes (2011) http://www.nfpa.org/research/reports-and-statistics/fire-causes. Accessed 30 Jan 2014
41. Vehicle Fires (2011) http://www.nfpa.org/research/reports-and-statistics/vehicle-fires. Accessed 30 Jan 2014
42. Marsh (2001) 100 Largest Losses 1972–2001. 100 Largest Losses
43. Marsh (2009) 100 Largest Losses, 1872–2009. 100 Largest Losses
44. Gugan K (1978) Unconfined vapour cloud explosions. UK ed. IChemE/Godwin (book)
45. Pemex LPG Terminal, Mexico City, Mexico Explosion, 19th November 1984. http://www.hse.gov.uk/comah/sragtech/casepemex84.htm
46. Explosion and fire at the Tesoro Refinery in Anacortes kills seven refinery workers on April 2, 2010. http://www.historylink.org/index.cfm?DisplayPage=output.cfm&file_id=9717. Accessed 03 Mar 2014
47. Explosion and fire at DSM, Beek. 7th November 1975. http://www.hse.gov.uk/comah/sragtech/casebeek75.htm
48. Lees FP, Chalmette P (1990) Louisiana explosion. In: Mannan S (ed) Lees' loss prevention in the process industries, vol 1
49. Chalmette (1976) LA oil refinery tower explosion. http://www3.gendisasters.com/louisiana/13110/chalmette-la-oil-refinery-tower-explosion-july-1976
50. Disasters and Major Accidents (Explosions). http://www.ilo.org/oshenc/part-vi/disasters-natural-and-technological/item/363-disasters-and-major-accidents
51. List of accidents and disasters by death toll. http://en.wikipedia.org/wiki/List_of_accidents_and_disasters_by_death_toll
52. UN Secretariat Committee of Experts on the Transport of Dangerous Goods and on the Globally Harmonized System of Classification and Labelling of Chemicals (2012) ST/SG/AC.10/C.3/2012/99. Accessed 30 Apr 14
53. List of accidents and disasters by death toll. http://www.survivalring.org/awareness/what-do-you-want-to-survive/accidents-disasters/
54. Bond J (1996) Close confinement. In: The hazards of life and all that. Institute of Physics Publishing

55. Gugan K (1978) Combustion and physical fundamentals. In: Unconfined vapour cloud explosions. IChemE/Godwin
56. Lees FP (2005) Explosion, in Lees' loss prevention in the process industries. In: Mannan S (ed) Elsevier, Butterworth Heineman, Oxford
57. Withers J (1988) Major industrial hazards: their appraisal and control. Gower Technical Press, Aldershot
58. Current World Population (2014) http://www.worldometers.info/world-population/. Accessed 31 Jan 2014
59. National Weed Strategy; Artichoke Thistle. http://www.weeds.org.au/cgi-bin/weedident.cgi?tpl=plant.tpl&state=&s=&ibra=all&card=H92. Accessed 19 Jan 2014
60. Galloway Wild Foods; Sea Radish; noxious weed in Australia [cited 20 Jan 2014; Schedule 4 weed in Australia]. http://www.gallowaywildfoods.com/?page_id=110
61. Weed risk assessment (2010) Horseradish tree—Moringa oleifera
62. Lincoln weed (Diplotaxis tenuifolia) Sand Rocket. http://archive.agric.wa.gov.au/PC_93097.html?s=1001. Accessed 28 Feb 2014

Design of a New Innovative High Pressure Hydrogen System Depending on Multi-layer Section

D. Duschek and J. Wellnitz

Abstract Based on the rising environmental pollution and finite fossil fuel new storage system must develop to dissipate the primary energy source of renewable energy to the second energy source, like hydrogen. A view into the field of pressure tank technology shows that usually today it is common practice to design a huge cylinder tank. On market, the state of the art high pressure storage systems, type III or type IV, are classified in the Regulation (EC) No 79/2009 [1]. The idea of the new design is to store as much as possible mass of hydrogen in a smaller volume without losing a lot of constructed space. The first idea is to reduce the inner diameter and create new design based on bending pipes. The smaller cross-sectional area of the pipe caused lower tangential and axial stresses with the result of reducing the wall thickness and resulting weight. This modification has a negative influence of the whole structure, because with decreasing diameter the relation between these significant masses also gets worse, as the inside volume of the pipes gets lower. The challenge of drafting a new high pressure storage system is to find a compromise between the optimum diameter for a good amount of inside volume and comparably low mass of the hydrogen storage structure. The comfort for the user, volume and mass of hydrogen for the car depends on the design.

1 Introduction

In times of rising availability of green electricity as primary energy source, new storage systems must develop to use renewable energy, because the source depends on environment influence, like sun or wind intensity and the amount of produced renewable energy is higher as consumed. In the future, the hydrogen storage system

D. Duschek (✉) · J. Wellnitz
Technische Hochschule Ingolstadt, Esplanade 10, 85049 Ingolstadt, Germany
e-mail: Daniel.duschek@thi.de

J. Wellnitz
e-mail: joerg.wellnitz@thi.de

I. Denbratt et al. (eds.), *Sustainable Automotive Technologies 2014*,
Lecture Notes in Mobility, DOI 10.1007/978-3-319-17999-5_20

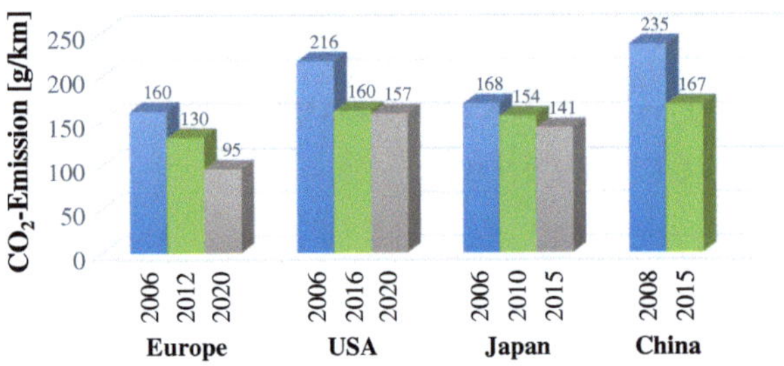

Fig. 1 Trend of CO_2 emission limitation depends on countries [2]

can be a solution to dissipate the energy as fuel used by mobility, like vehicle or to heat buildings. Another aspects are the growing environment pollution and finite source of fossil fuel. Especially the automotive industry must find a solution, because the CO_2 emission limit of countries lead to upgrade the state of the art powertrains of modern vehicle. The specifications depending on environment rules shown in Fig. 1 [2]. The comparison of Europe and China illustrate that European limit of 160 g/km in 2006 is lower as the future aims 167 g/km of the year 2015 in China. The new storage system can used for other sector, too. Actually, the increasing trend of civil aviation caused environment pollution and the market demand of fossil fuel next to automotive industry [3]. Therefore, the new design based on modular construction and easily modified to using hydrogen for aircraft [4] or as mobile power station, too.

The medium hydrogen has from all fuels the highest mass-specific energy density. A comparison of the different available storage strategies shows the following kind of storage systems:

- gaseous
- liquid
- solid

The gaseous storage technology has established in the 19th century [5] and since then the preferred system, because the cylinder shape can easily manufacture and the dimensioning can calculate by reference to DIN 2413 [6]. The knowledge are the basis of testing hydrogen as fuel in the field of aircraft, e.g. B-57 or TU 154. In history, the European studies like **Cryoplane**, **Consove 2050** and **storhy** shows the potential of the new technologies. Nowadays, Boeing, EADS and the DLR, e.g. Antares H3, research hydrogen as alternative fuel. The application to vehicle, aircraft or buildings is possible, because the technique is well known. The parameter depends on the influence are controllable and the mass production can realized. The structure of the tanks consist of an inner diffusion resistant aluminium layer and a carbon composite outside liner with the purpose of carrying the pressure loads. The American Department of Energy (DOE) published future targets, which describe the system parameter of new hydrogen storage systems [7]. The benchmark analyse of available tanks shows that the relation of gravimetric and volumetric density have to optimize. The aims can reach by raising up the pressure level from 35 to 70 MPa. The result is higher stress and more weight of the storage system. The idea is to design a new compressed gas storage tank with reducing the diameter, e.g. d = 30 mm and develop new shape like meander or helix. The tank can be a part of the support structure, for example as frame into b-pillar of a vehicle or airplane wing as result of the smaller dimensioning.

2 Structure of the New Hydrogen Storage System

The challenge of the design process is to find the compromise between used constructed size, capacity and weight of gaseous storage systems. The analysing of Type III hydrogen storage system, categorized EG79/2009 [1], based on vessel geometry. To adapt the tank as supported structure the design can have a freeform shape, e.g. spline. Next to cylinder it's important to have a view on circular ring (see Fig. 2).

The axial stresses σ_a of a straight pipe and a torus are similar (see Eq. 2.1).

$$\sigma_a = \frac{F_t}{A_t} = \frac{\Delta p d_i}{4t} \tag{2.1}$$

The tangential stresses σ_t of a circular ring depends on the section radius r_m of the circle, the radius angle nominal to bending radius R_k (see Eq. 2.2) [8].

$$\frac{\sigma_t}{r_m} + \frac{\sigma_a \sin\eta}{R_k + x} = \frac{\Delta p}{t} \tag{2.2}$$

With insert $x = r_m \sin\eta$ in the Eq. 2.2 and assumptions following to calculate the torus depending on the angle η (see Eq. 2.3).

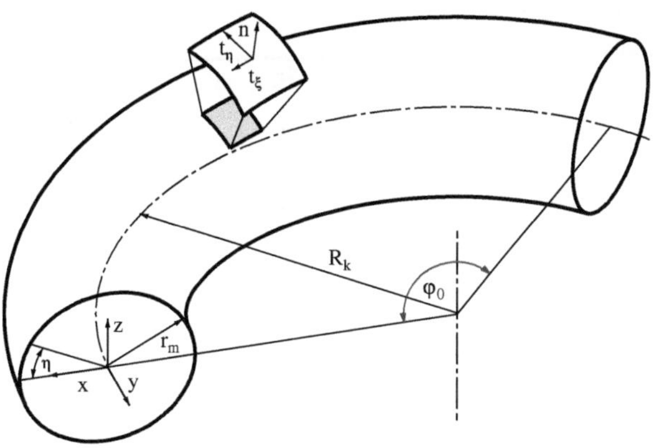

Fig. 2 Parameter of Torus geometry with constant circle section

$$\frac{\sigma_t}{r_m} + \frac{\Delta p}{2t} \frac{r_m \sin \eta}{R_k + r_m \sin \eta} = \frac{\Delta p}{t} \quad (2.3)$$

and therefore

$$\Rightarrow \sigma_t(\eta) = \sigma_\eta = \frac{\Delta p d_m}{2t} \frac{\frac{R_k}{d_m} + 0.25 \sin \eta}{\frac{R_k}{d_m} + 0.5 \sin \eta} \quad (2.4)$$

The Eq. 2.4 shows that $\sigma_\eta(\eta = 90°)$ is smaller as $\sigma_\eta(\eta = 0°)$, but greater than $\sigma_\eta(\eta = -90°)$. The result of the first calculation is, the stress of a torus in axial direction are similar to the cylinder, but the tangential stresses caused by the pressure level are higher (see Fig. 3).

The result of design tank structure clarify, that the stresses of torus shape caused by internal pressure define the wall thickness. The benchmark of state of the art compressed gaseous tanks of type I are too heavy to reach the future target, e.g. DOE of the year 2017. The idea is additionally reducing the mass based on composite structure. The hybrid structure seen on Fig. 4 is that the metal liner (1) define the shape and used as barrier coat of the medium hydrogen.

The extruded aluminium sheath is the raw material to produce bending pipes. The following number of layers (2), (3) … (n) based on fibre reinforced plastic and depends on the pressure level. The fibre absorb the stresses caused by internal pressure Δp and the matrix material, e.g. Epoxy locate the fibre angle. Figure 5 shows the results of the parameter study by inserting the fibre and epoxy material parameter, volume fraction and safety factor into the Eq. (2.4) and solving to whole wall thickness t_{sum} for 70 MPa internal pressure. The wall thickness of the aluminium liner is constant with $t_{Al} = 1$ mm.

A view on the calculation of the diameter $d_i = 120$ mm and the fibre material Innegra clarify, that a whole wall thickness $t_{sum} = 30$ mm to realize is very expensive, because many composite layers are needed. Following, the

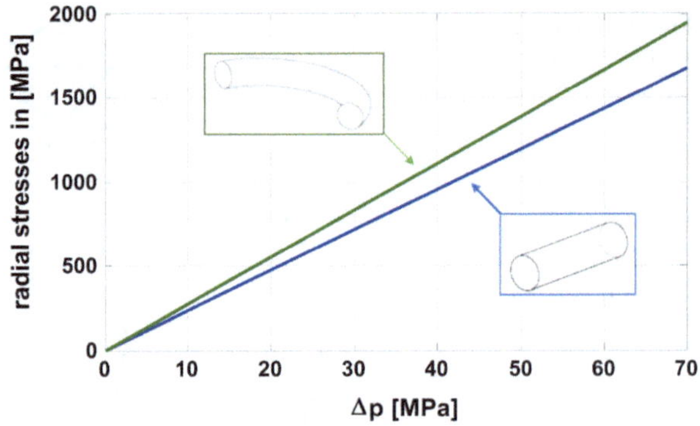

Fig. 3 Comparison of tangential stresses σ_t of straight pipe to bending pipe depending on internal pressure

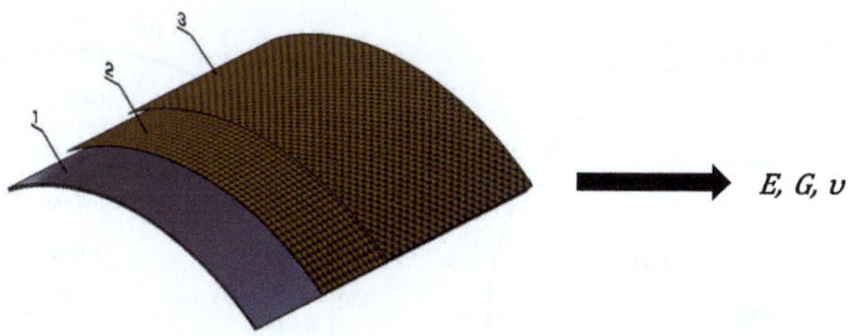

Fig. 4 Assembly of the hybrid layer structure

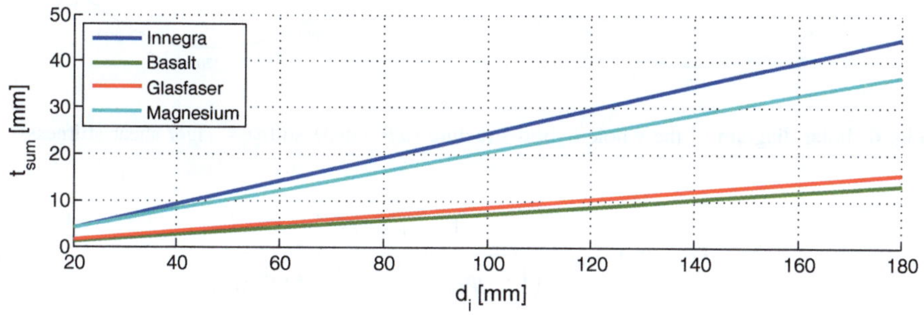

Fig. 5 Wall thickness t_{sum} depending on diameter d_i and internal pressure Δp

manufacturing costs can reduce by less number of composite layers. Another point of view can be the production of a completely metal part. This mean, the specification on the material must be low density and high tensile strength. The properties of Magnesium, e.g. AZ61 are very good and the aim of low mass can reached. However, in the case of d_i = 120 mm the wall thickness is nearly 25 mm. The basalt fibre is the best solution with a total material thickness of 8 mm. The properties of basalt, e.g. available on market and tensile strength are suitable conditions for the hybrid tank structure.

The behaviour of fibre material are anisotropic. Therefore, the load level depends on fibre orientation direction. The hybrid based on a defined number unidirectional (UD) single layer of fibre reinforced plastic and a constant metal liner of aluminium. The derivation to calculate the stress each layer according to "world-wide-failure-exercise" of Schürmann and Puck [9] as following for the semi-empirical equation of young's modulus $E\perp$, Eq. 2.5

$$E_{\perp} = \frac{E_{resin}}{(1 - \nu_{resin}^2)} \cdot \frac{1 + 0.85 \cdot \varphi_{Comp}^2}{(1 - \varphi_{Comp})^{1,25} + \frac{E_{resin}}{(1 - \nu_{resin}^2) \cdot E_{fibre\perp}} \cdot \varphi_{Comp}} \tag{2.5}$$

and the shear modulus $G_{\perp\parallel}$, Eq. 2.6

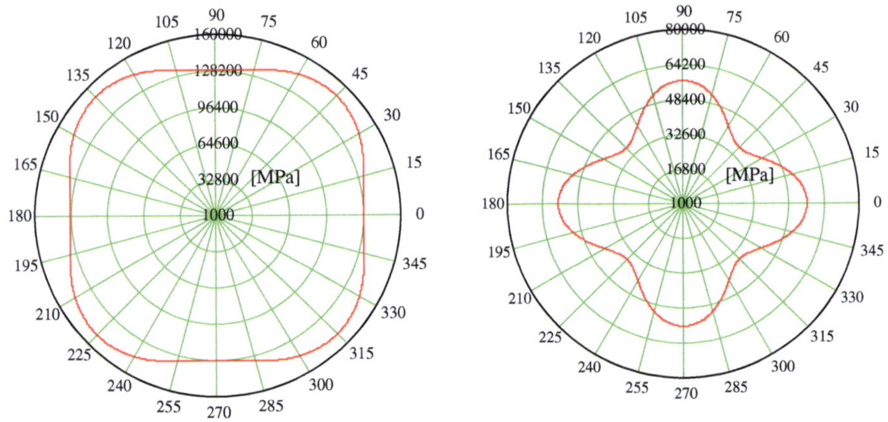

Fig. 6 Polar diagram of the whole hybrid structure (*left* tensile stiffness, *right* shear stiffness)

$$G_{\perp||} = G_{resin} \cdot \frac{1 + 0.4 \cdot \varphi_{Comp}^{0.5}}{\left(1 - \varphi_{Comp}\right)^{1.45} + \frac{G_{resin}}{G_{fibre}} \cdot \varphi_{Comp}} \tag{2.6}$$

are closed to test results.

It is necessary to set up the stiffness matrix with the classical laminate theory [10], because the different mechanical properties of used materials, like fibre direction each layer are include. To calculate the sum stiffness matrix of the multilayer composite the coordinate system of the fibre orientation each UD single layer transformed to the polar base (x, φ, r) and summarize by principle of superposition. The result of the calculation with Δp = 70 MPa, basalt fibre, fiber volume content of 50 % shown in Fig. 6. The polar diagram depends on the inner diameter d_i of the pipe, number of layer, young modulus, shear modulus, orientation.

The simulation of the multilayer composite with metal liner clarify that the behaviour of a hybrid structure is practical, because the trend curve of tensile stiffness on the polar diagram smoothed at 0°, 90°, 180° and 270°. In comparison at 90°, the tensile stiffness of the hybrid structure is 128 and 50 GPa of the composite structure.

3 Conclusion/Summary

The scientific findings of the parameter study and theoretical view are the base of simulation program. The aim is to calculate the new hydrogen storage system depending on design, different construction space, weight of tank and storage volume. The modular design can used as stationary tank or as application in the field of vehicle or airplane. The challenge is to find the best conclusion between all facts and define the system parameter, because the assembly of the design is similar to each dimension. The analyse of different size shows, that one of the most important system value is the inner diameter, because d_i has influence on curvature radius, storage volume, construction space and a lot of more points. The knowledge of prototype manufacturing can used to revise the design.

References

1. Regulation (EC) no 79/2009 of the European Parliament and of the Council of 14 January 2009 on type-approval of hydrogen-powered motor vehicles, and amending Directive 2007/46/EC
2. Schlick T (2013) Zukunft der Elektromobilität. auto.mobilFORUM Veritas Gelnhausen, November 2013
3. Mensen H (2013) Handbuch der Luftfahrt. Springer, New York
4. Böhm M, Bundeswehr München, Universität der (Hrsg.) (2007) Gesamtentwurf eines ökonomischen und ökologischen Lufttransportsystems unter Ausnutzung von Synergieeekten. Dissertation
5. Novak P (2006) Ein Beitrag zur Strukturoptimierung dickwandiger Hybrid-Hochdruckspeicher. VDI-Verlag
6. DIN 2413 (2010) Seamless steel tubes for oil- and water-hydraulic systems—calculation rules for pipes and elbows for dynamic loads
7. DOE; 4, Revision: targets for onboard hydrogen storage systems for light -duty vehicles. US Department of Energy Office of Energy Efficiency and Renewable Energy and The Freedom CAR and Fuel Partnership, September 2009
8. Annaratone D (2007) Pressure vessel design. Springer, Berlin
9. Schürmann H, 2., bearb. und erw. A. (Hrsg.) (2007) Konstruieren mit Faser- Kunststoff-Verbunden. 2. Springer, Berlin
10. Basan R (2011) Untersuchung der intralaminaren Schubeigenschaften von Faserverbundwerkstoffen mit Epoxidharzmatrix unter Berücksichtigung nichtlinearer Effekte Bundesanstalt für Materialforschung und -prüfung (BAM)

Part VII
Safety

Emergency Management Support by Spatial Reasoning

Jan Olaf Blech, Heinz Schmidt and Timos Sellis

Abstract Emergency management benefits from techniques such as the eCall that allow an automatic transmission of vehicle data and location to initiate response operations in case of an accident. Such operations may comprise the deployment of ambulances and recovery vehicles. Based on available data one can decide on the type of ambulances, police and other recovery vehicles needed, on prioritization in case of multiple events and on strategies for an efficient management of available resources. Automatically handling these constraints which can go beyond traditional database operations and deriving decisions is a challenging problem. In this paper, we describe how our existing spatio-temporal description and reasoning framework based on formal methods can be used to facilitate decisions in emergency recovery situations in combination with indexing of available information.

1 Introduction

Modeling techniques for software and systems have been widely studied in the past including models for the environments in which a system is or will operate. Likewise, tools supporting simulation, verification/checking of properties and validation of models have been developed for domain specific purposes. Here, we are studying spatial behavioral models of entities such as ambulances and helicopters for emergency management. Questions regarding sustainability in particular the efficient use of resources appear at the level of emergency response fleet management, and for adequate responses to incidents including resolving possible sources of disruption (e.g., accidents causing traffic-jams). These appear as additional constraints in our models. We are using our BeSpaceD spatial verification tool [3, 4] and combine it with ideas from geospatial databases to reason about scenarios that go beyond traditional database queries. We reformulate decision and

J.O. Blech (✉) · H. Schmidt · T. Sellis
RMIT University, Melbourne, Australia
e-mail: Janolaf.Blech@rmit.edu.au

I. Denbratt et al. (eds.), *Sustainable Automotive Technologies 2014*,
Lecture Notes in Mobility, DOI 10.1007/978-3-319-17999-5_21

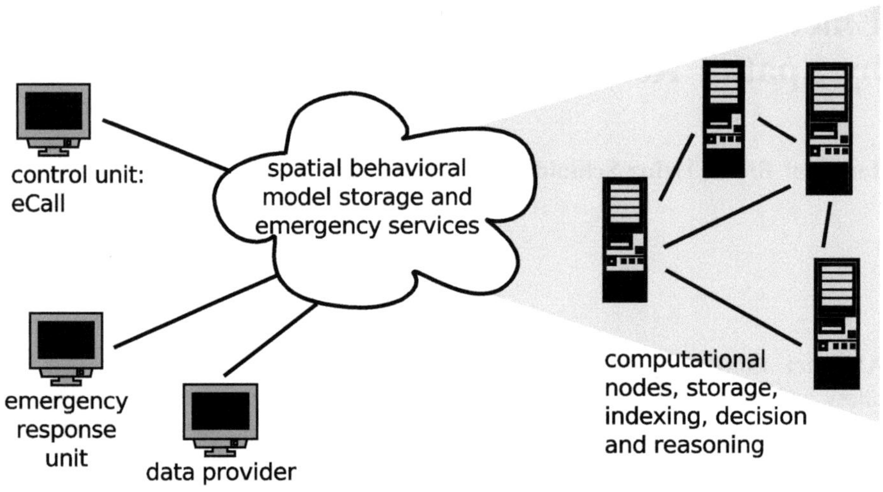

Fig. 1 Spatial models for emergency handling in the cloud

optimization questions in emergency handling as formal methods verification problems. Figure 1 gives an overview of a proposed architectural plan for handling spatial models in the cloud. Storage and emergency response decisions using spatial behavioral models is done in a parallel computing environment in the cloud. In addition to a model indexing infrastructure that keeps track of parts of a model, the parallel computing environment offers computational nodes on which different parts and abstract views of a model can be stored and used for emergency response decisions. Storage and emergency response decisions are offered as a service. These services may be provided in a standardized way as web-services and can be accessed not only from end-users directly (e.g., via a web-interface or out of a control unit in a car), but also from tools (e.g., plug-ins for the Eclipse software development environment) that run on user machines.

Spatial behavioral models used in this work are related to description logics of spatial systems (see [5, 6] for a process algebra like formalism).

2 Motivating Example

The eCall initiative[1] aims at providing automatic emergency calls for vehicles involved in accidents to speed up rescue operations. These calls include geospatial information which can be handled automatically. The initiative has developed

[1]http://ec.europa.eu/information_society/activities/esafety/ecall/index_en.htm.

a standard that is now widely adapted throughout the automotive industry. Another approach is the Israeli *United Hatzalah* project.[2] It aims at speeding emergency services up by coordinating volunteer motorcycle-based ambulances based on their geospatial positions and availability. This also requires the retrieval and handling of spatial information. Based on these projects, here, we regard spatial models, where participants in emergency rescue operations (like ambulances, medical helicopters, technical rescue workers)

1. can store more *course grained approximations* of their expected spatial roaming behavior and capabilities and update it occasionally, and
2. use more, *fine grained updates for local changes* that we can keep track off if available in addition to the information provided by 1.

Both information from (1) and (2) can be stored in the same model, but different views on the model limit the access to information.

Similar to rescuers, participants who may be in need for rescue operations can store their expected behavior. In case of an emergency, we retrieve the spatial coordinates of the emergency site and match it with activity areas or—if available—exact coordination and activity status of potential rescuers. The handling involves the interpretation of different models and data, since multiple rescuers may be required or combinations of rescue devices, each one in possession of a particular tool or skill may be needed to conduct the rescue operations. An emergency situation is depicted in Fig. 2. An area with an accident is shown. Planes indicating the estimated availability of ambulances and helicopters are also shown. One can imagine, that each ambulance and helicopter comes with different spatial data indicating likely availability for geographical locations and time (not shown in the Figure). For ambulances that are stationed at particular locations, the time aspect will not change frequently, while individuals providing first-aid assistance may have a much more unpredictable spatial behavior.

In our work, we assume that we can associate time intervals with actors (e.g., ambulances), their characteristics (e.g., special equipment) and different spatial areas indicating likely availability or explicitly stating a probability.

Based on this, we can

- Check—verify using BeSpaceD—if and whether enough help will be available for a distinct accident and contact the involved participants based on both the fine grained and the course grained information.
- Similarly we may be interested in checking more general safety and consistency related questions, like, if an accident would occur in a certain area, what is the response time that we can guarantee? These questions will be verified based on the coarse grained model view.

[2] http://unitedhatzalah.org.

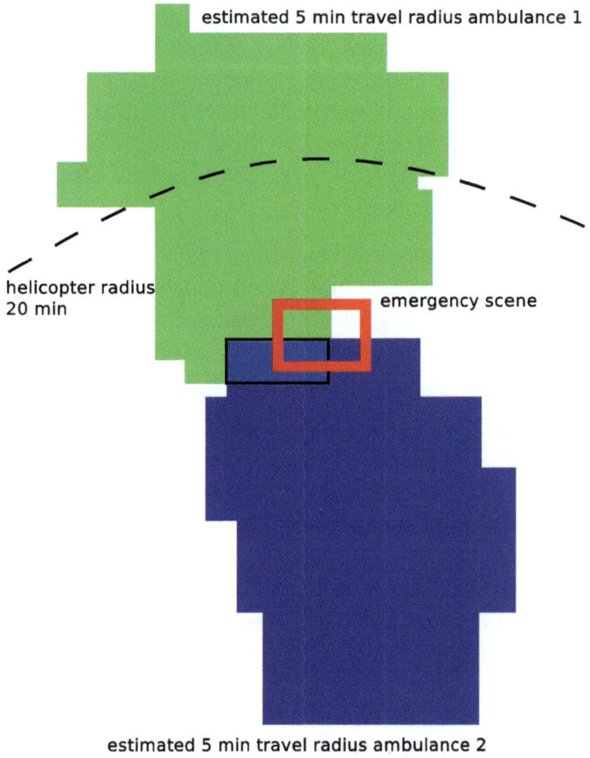

Fig. 2 Emergency situation

3 Framework

Specification of widely distributed system models can be done using different formalisms such as automata with annotated spatial descriptions and architecture description languages including UML diagrams (e.g., extended behavioral types [2]), geospatial models and combinations of different specification languages especially for software and spatial parts of a model. Our BeSpaceD modeling language allows the time or automata based behavioral descriptions for entities in combination with their spatial (coordinates or topological) characteristics. These can be used to describe availability areas and schedules, capabilities and additional aspects such as events and states. For example, an availability area can vary depending on the time and date and external events such as traffic jams. A simplified example description of a spatial behavioral model is provided below. It consists of logical formulae expressing constraints on time, space, states, events and probabilities.

```
BIGAND(
  IMPLIES(AND(TimePoint(...), Event (...)),OccupySpace(...)),
  IMPLIES(AND(TimeInterval(...), State (...)),OccupySpace(...)),
          ...
  IMPLIES(AND(TimePoint(...), Event (...)),OccupySpace(...)),
  IMPLIES(AND(TimeInterval(...), Probability (...)),OccupySpace(...)))
```

Updates to spatial models can occur from the different participants and on different scales, for example

- Adding new actors, e.g., new ambulances does not occur very often. Each ambulance comes with a rough specification of its deployment area and availability schedule.
- Changing the deployment area and a rough plan of availability of an actor may be known several weeks in advance. For example, maintenance schedules may be taken into account, as well as assigning an ambulance to another hospital.
- Keeping track of a driving ambulance in a local area happens on a much finer time and space scale (e.g., an update per minute or even second).

4 From Models to Spatial Properties Checking

Figure 3 gives an overview on our proposed checking infrastructure. Models are abstracted and the abstractions are indexed using database techniques (e.g., k-d tree [1] based indexing with respect to time and space). Finer-update information that may arrive on a minutely bases can be stored separately from the original models and accessed using the same indexing techniques.

Our optimization and decision questions, in conjunction with our spatial behavioral models are given to our tool framework, the database based indexing (cf. e.g., [9]) is used to recover all important entities and their models/information for an area

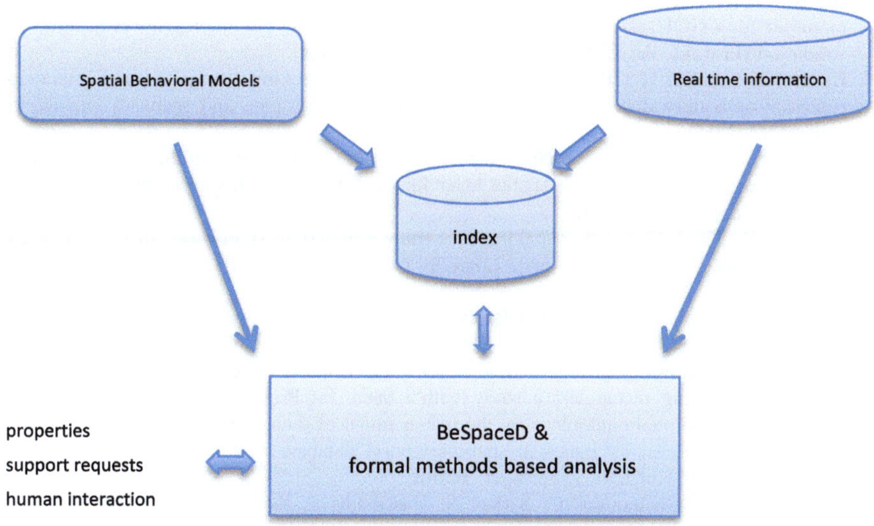

Fig. 3 Architectural overview

and time. Our framework breaks this information down to expressions containing geometrical information. Special algorithms on analyzing spatial relationships can be used to evaluate these expressions [7, 8, 10]. Another automatic transformation breaks these geometric expressions down to representations that are suitable for solving algorithms and special solvers such as SMT and SAT based techniques for which different implementations exist in the formal methods community.

5 Conclusion

We have motivated and presented work for using our spatial verification framework together with data based indexing for supporting emergency recovery operations. Our framework is able to decide and optimize recovery operations based on formal spatial behavioral models and formal methods. In addition to our existing work on spatial reasoning and geospatial databases, we have listed implications and sketched an architecture for this new application domain. An end-user may eventually be able to use these services automatically (triggered), via a web-interface or out of customized application and devices. Likewise other frameworks can use them using SOA style mechanisms such as web-services.

References

1. Bentley JL (1975) Multidimensional binary search trees used for associative searching. Commun ACM. 18 (9):509–517
2. Blech JO, Falcone Y, Rueß H, Schätz B (2012) Behavioral specification based runtime monitors for OSGI services. Leveraging applications of formal methods, verification and validation (ISoLA), vol 7609 of LNCS. Springer, New York
3. Blech JO, Schmidt H (2013) Towards modeling and checking the spatial and interaction behavior of widely distributed systems. In: Improving systems and software engineering conference, Melbourne
4. Blech JO, Schmidt H (2014) BeSpaceD: towards a tool framework and methodology for the specification and verification of spatial behavior of distributed software component systems. http://arxiv.org/abs/1404.3537
5. Caires L, Cardelli L (2003) A spatial logic for concurrency (Part I). Inf Comput 186/2:194–235
6. Caires L, Cardelli L (2004) A Spatial Logic for Concurrency (Part II). Theor Comput Sci 322 (3):517–565
7. Papadias D, Sellis T (1994) Qualitative representation of spatial knowledge in two-dimensional space. VLDB J 3(3):479–516
8. Papadias D, Theodoridis Y, Sellis T, Egenhofer M (1995) Topological relations in the world of minimum bounding rectangles: a study with r-trees. In: Proceedings of the 1995 ACM-SIGMOD international conference on the management of data, San Jose, pp 92–103
9. Sellis T (1999) Research issues in spatio-temporal database systems. Advances in spatial databases, vol 1651 of LNCS. Springer, Berlin
10. Skiadopoulos S, Giannoukos C, Sarkas N, Vassiliadis P, Sellis T, Koubarakis M (2005) Computing and managing cardinal direction relations. IEEE Trans Knowl Data Eng 17 (12):1610–1623

Analytical and Numerical Approach of Dynamic Behaviour of Flexible Metal Mesh Structures

E. Wilhelm and J. Wellnitz

Abstract The application of Flexible Metal Mesh Structures (FMMS) has the potential to improve human and vehicle safety. Previous experiments confirmed the effectiveness of these structures. Experimental tests are time and cost consuming. Thus the aim of this study is to develop an appropriate simulation model, due to economical and timed aspects. Therefore, within this study an analytical and numerical model at macro-level will be established simulating the dynamic behaviour of the base chainmesh through defined parameters and boundary conditions. The main focus of the present paper is to understand the kinematic behaviour of chainmesh under large-scale impact in the out of plane direction of the material. The field of future applications (human and vehicle safety) defines the governing parameters of the model. Accordingly, high loads and slow impact velocities are simulated, while small distortion and large deformations within the material are tolerated. Furthermore, to ensure reusability of the metal mesh, plastic deformation should not occur, thus have not to be captured within the model. In view of its future applications, the following key topics will be investigated:

- DOFs and kinematics of the single (torus) element,
- motion of the overall system,
- energy absorption of the chainmesh and
- estimation of risk of injury in case of human body impact.

Thus, the main aim will be the applicability of the model with its practical relevance lying in the potential to improve Flexible Metal Mesh Structure applications in the field of human and vehicle safety.

E. Wilhelm (✉)
Institut für Technik und Design GmbH, Nördliche Ringstr. 19, 85057 Ingolstadt, Germany
e-mail: eva.wilhelm@itd-in.de

J. Wellnitz
Technische Hochschule Ingolstadt, Esplanade 10, 85055 Ingolstadt, Germany
e-mail: joerg.wellnitz@thi.de

I. Denbratt et al. (eds.), *Sustainable Automotive Technologies 2014*,
Lecture Notes in Mobility, DOI 10.1007/978-3-319-17999-5_22

1 Motivation

Flexible Metal Mesh Structures (FMMS) are used as a protection system, where humans, vehicles or machine parts are exposed to hazards. The major advantages of this particular protection system are adaptable mechanical properties, retrofitting capacity, application variety and transparency, proven by various large scale and local impact tests as well as dynamic tests [2, 5–7].

However, improvements within the protection sector are required, because flexibility and adjustability is highly important for military and civilian environments to be retrofitted, mentioned in [3].

To reflect such an event of an impact on a numerical basis an appropriate simulation model is necessary, which hasn't been developed for a chainmesh application yet. So the main aim of this work is to understand the dynamic behaviour of chainmesh during an impact and transfer this knowledge in an appropriate numerical model. Therefore, the chainmesh has to be analysed analytically within the macro-level first. Besides the documentary in [4] and [8] only one thesis [5] about this special topic has been published so far, where the location of the contact points and the interaction of contact forces were described as a reaction of an in-plane tensile load. Due to symmetry effects the reference for this analysis was one-quarter of a ring element. However, the implementation of the analytical data within a numerical simulation requires the sum of all responding forces at a single ring not only at one-quarter. Therefore, the analytical investigation needs improvement in order to validate a realistical numerical representation.

2 Methodology

FMMS describes a chain meshwork, which consists of a large number of welded metal rings with a well-defined geometry. The mesh used in this study is a 4 in 1 configuration, which means that every single ring is connected with four neighbouring rings (see Fig. 1).

Fig. 1 Detailed geometric view of the 4:1 chainmesh configuration and the resulting contact points *K1–K8* between the central ring (*Z1*) and the four connected rings (*R1–R4*) and *K9* and *K10* between the central ring and the two neighbouring rings

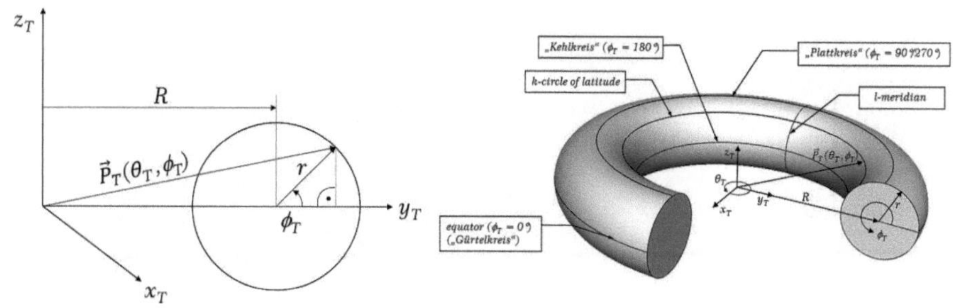

Fig. 2 Generation of a torus by revolving a circle with radius r around the z-axis with a certain distance R (*left*); general topology and terminology of a torus element (*right*), own representation based on [5]

2.1 Analytical Approach

In analytical terms each single ring can be idealised as a torus according to several tension test results, where it was shown that the failure of the mesh takes place in the narrowing of single rings independent of the weld seam placement. A torus is a rotationally symmetrical body with a closed surface, which results in revolving a circle with radius r around the z-axis in a certain distance R (Fig. 2).

Every single point on the surface of the torus can be described via a position vector. The result of the parameterization of a torus within the z_T-Y_T-plane where the angles θ_T and ϕ_T are defined as $0 \leq \theta_T, \phi_T \leq 2\pi$ is:

$$\vec{P_T} = \begin{pmatrix} -\sin\theta_T(R + r\cdot\cos\phi_T) \\ \cos\theta_T(R + r\cdot\cos\phi_T) \\ r\cdot\sin\phi_T \end{pmatrix}$$

Furthermore can the metric of the torus be deduced from the partial derivation of the position vector $\vec{P_T}$ with the Cartesian coordinate system placed in the centre of the torus. If the coordinate system is placed somewhere else translational and rotational components of the torus position has to be considered.

2.2 Numerical Approach

Parallel to the analytical work a software research was done within this study, which pointed out that the software LS-DYNA is able to represent the desired simulation model. With it, explicit contact-impact interfaces and further contact characteristics can be simulated.

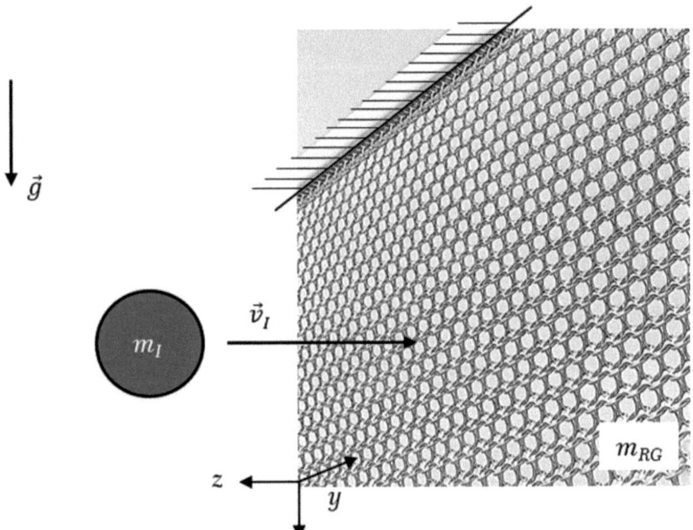

Fig. 3 A vertical hanging arrangement of chainmesh as a result of the dead weight m_{RG} of the mesh, which is suspended by an one-sided fixation at the *top*; mass m_I impacts the chainmesh according to Cartesian coordinate system in *z*-direction with a velocity v_I

The simulation model cannot cover all possible course of events and represent prior ideal conditions instead. Figure 3 shows an exemplary situation, where an impact mass m_{RG} reaches the chainmesh with mass m_I in the out of plane direction (*z*-direction) of the material.

In this study high loads and slow impact velocities of the impact mass m_I are decisive initial conditions. In addition, the main parameters and the boundary conditions are defined as following:

- tolerance of small distortion and large deformations within the material
- ensurance of reusability of the material
- plastic deformation will not be captured in the model

Besides LS-DYNA other representative possibilities were proven. It is believed that a reproduction of the stable equilibrium position of the chainmesh can be realized according to the Lennard-Jones potential, see also [1]. At this state of study it is unknown if the representation of the dynamic behaviour and further adjustments in size and impact characteristics can be also covered with the potential theory approach.

Both mentioned approaches suit the initial requirements of including all contact points and contact forces of a representative chainmesh section in order to obtain a more realistic simulation.

3 Conclusion/Summary

Due to several experimental studies it has been proven that as a reaction of the impact the one-sided suspended chainmesh wraps itself around the impact mass, which leads to an energy absorption and a decrease of velocity. These tests verify especially the potential of Flexible Metal Mesh Structures (FMMS) concerning protective effects and substantial potentials for debris containment within several speed and impact mass regimes.

Furthermore, this study implements the existing experimental results and the improved analytical data in a new modelling approach in order to validate a realistical numerical representation of the chainmesh behaviour during an impact.

The next steps will include a mathematical description of all contact points with regard to one exemplary central ring (see Fig. 1) and the resulting contact forces. This knowledge and the kinematic behaviour is needed to transfer the mechanical behaviour to a chosen software programme, where the kinetics can be analysed in detail.

References

1. Balke H (2010) Einführung in die Technische Mechanik. Springer, Berlin
2. Burger U (2009) Impactverhalten von hybriden Verbundwerkstoffen mit metallischem Ringgeflecht. Dissertation, Technische Universität Berlin
3. Gebbeken N, Döge T (2010) Explosion protection—architectural design, urban planning and landscape planning. Int J Prot Struct 1:1–22
4. Hahn M, Bouriga M, Kröplin B-H, Wallmersperger T (2013) Life time prediction of metallic materials with the discrete-element-method. Comput Mater Sci 71:146–156
5. Lenk O (2009) Charakterisierung und Anwendung von Metall-Ringgeflechten. Dissertation, Brandenburgische Technische Universität Cottbus
6. Wilhelm E, Burger U, Wellnitz J (2012) Investigation of chain meshwork: protective effects and applications. In: Proceedings of the 4th international conference on sustainable automotive technologies
7. Wilhelm E, Wellnitz J (2013) The usage of lightweight materials in hazardous areas: flex-metal-mesh. In: Proceedings of the 5th international conference on sustainable automotive technologies
8. Zhodi TI, Mseis GN (2011) Micromechanical modeling and numerical simulation of chain-mail armor. Springer Science + Business Media, New York

MIX
Papier aus verantwortungsvollen Quellen
Paper from responsible sources
FSC® C105338

If you have any concerns about our products,
you can contact us on
ProductSafety@springernature.com

In case Publisher is established outside the EU,
the EU authorized representative is:
Springer Nature Customer Service Center GmbH
Europaplatz 3, 69115 Heidelberg, Germany

Printed by Libri Plureos GmbH
in Hamburg, Germany